Vinicius Armando Covielo
Daniel Andrade

Citrus leprosis: Fruit characteristics x disease severity

Vinicius Armando Covielo
Daniel Andrade

Citrus leprosis: Fruit characteristics x disease severity

Chemical and physical results of pear orange fruit analysis as a function of disease severity

ScienciaScripts

Imprint
Any brand names and product names mentioned in this book are subject to trademark, brand or patent protection and are trademarks or registered trademarks of their respective holders. The use of brand names, product names, common names, trade names, product descriptions etc. even without a particular marking in this work is in no way to be construed to mean that such names may be regarded as unrestricted in respect of trademark and brand protection legislation and could thus be used by anyone.

Cover image: www.ingimage.com

This book is a translation from the original published under ISBN 978-3-330-75639-7.

Publisher:
Sciencia Scripts
is a trademark of
Dodo Books Indian Ocean Ltd. and OmniScriptum S.R.L publishing group

120 High Road, East Finchley, London, N2 9ED, United Kingdom
Str. Armeneasca 28/1, office 1, Chisinau MD-2012, Republic of Moldova, Europe
Printed at: see last page
ISBN: 978-620-8-27789-5

SUMMARY

1 INTRODUCTION

The citrus leprosis mite *Brevipalpus phoenicis* (Geijskes, 1939) (Acari: Tenuipalpidae) is one of the main pests of citrus farming as it is the vector of the CiLV (*Citrus* leprosis virus) which causes citrus leprosis. Leprosis is one of the most serious citrus diseases in Brazil, mainly due to the expense required to control the mite-vector (BASTIANEL et al., 2010).

Leprosis causes serious damage to citrus production, especially in the state of Sao Paulo, which is responsible for approximately 80% of national orange production (AGRIANUAL, 2009), giving the country the *status* of the world's largest citrus producer and largest exporter of concentrated and frozen juice.

Plants attacked by leprosis have a reduced photosynthetic capacity, the branches dry out and there is severe defoliation. When there is a severe incidence of the disease, the orchard can become economically unviable due to low productivity (CHIAVEGATO et al., 1982; OLIVEIRA et al., 1991). Controlling the mite *B. phoenicis* with acaricide applications has been the main leprosis management tactic and is responsible for a significant portion of production costs (BASSANEZI, 2004).

However, Chiavegato (1995) reported that the importance of leprosis in Brazil cannot only be assessed because of its widespread occurrence and the serious damage it causes, but also because of the difficulties in managing it, which require specific knowledge about the relationships between the vector, virus, host plant and environmental factors.

The causal agent of leprosis was identified by Kitajima et al. (1972), who observed bacilliform viral particles approximately 40 nm in diameter and 100-120 nm long inside the nucleus in sections of orange leaves with leprosis symptoms under a transmission electron microscope. In 1974, Kitajima et al. found viral particles in the lumen of the endoplasmic reticulum and dense viroplasms in the cytoplasm, identifying two distinct types of citrus leprosis virus, the cytoplasmic (CiLV-C) and the nuclear (CiLV-N) (RODRIGUES et al., 2003).

In Brazil, the transmission of leprosis in field conditions is attributed exclusively to the *B. phoenicis* mite which, when feeding on infected tissue cells, ingests viral particles and subsequently infects healthy tissues when it feeds again. However, there is no transovarian transmission of the virus, which is why mite dispersal plays a fundamental role in the evolution of the disease in citrus groves (RODRIGUES et al., 1997).

As a disease that has deleterious effects on plants, various physical and biochemical alterations can occur in the host plant, such as a reduction in photosynthesis, an increase in respiration and alterations in the synthesis of proteins and amino acids, among other effects (VICENTE, 1990). In addition, infections caused by diseases, especially those caused by viruses, due to their greater specificity with the plant, generally have significant effects on plant metabolism (HULL, 2002; FREITAS-ASTÙA et al., 2007), and can accelerate or delay fruit ripening, as well as altering their physical and chemical characteristics. Several studies have recently been carried out with the aim of explaining the main changes observed in sweet oranges infected with CiLV (FREITAS-ASTÙA et al., 2007).

In view of this, citrus leprosis, due to its uniqueness and the multiplicity of factors involved, as well as the complexity of the existing interactions, is an extremely complex pathosystem (BASSANEZI, 2001). This makes it necessary to carry out studies aimed at better understanding the effects caused, the variables involved and the interactions between them, in order to increase the efficiency of the control strategies employed (RODRIGUES et al., 2001).

The aim of this work was to study some of the physico-chemical characteristics of fruit in Pera citrus plants with different levels of leprosis severity.

2. LITERATURE REVIEW

2.1 Citrus leprosis

In 1907, the first record of citrus leprosis was made by Fawcett, in the Pinellas district of Florida, USA, called "scaly bark". In 1911, the disease reached its peak in that country, affecting around 75% of the orchards in 17 citrus-producing towns in the central region of Florida, and was called "nail - head rust". Later, in the 1930s, through photographic comparisons of the symptoms of "scaly bark" and "nail head hust", with the "explosive leprosy" of Argentina and the citrus leprosis of Brazil, Fawcett and Knorr concluded that they were the same disease. In the USA, the disease was gradually brought under control and by the mid-1950s it had disappeared, thanks to adverse climatic conditions and rigorous crop management, which included heavy applications of sulphur, which drastically reduced the population of the vector mite, which according to Knorr (1968), was *Brevipalpus californicus* (Banks, 1904) (CHILDERS et al., 2003b).

In South America, reports of the disease began to appear in the 1920s, under the name "explosive leprosy", which affected orange plants in Missiones, Argentina (SPEGAZZINI, 1920). In Brazil, the first report was made by Bitancourt (1937) who found the disease in the state of Sao Paulo in the 1930s, calling it citrus leprosis. Reports of its occurrence in Asia and Africa have not been confirmed (MURAYAMA et al., 2000) and there are no records of the disease in Europe, which shows that leprosis is restricted to the Americas.

Leprosis is more prevalent in sweet oranges [*Citrus sinensis* (*L.)* Osbeck], oranges (*C. aurantium*) and pomelos (*C. paradisi*), and is less prevalent in other citrus species, such as mandarins (*C. reshni Tanaka, C. deliciosa Tenore, C. reticulata*), hybrids such as Tangor Murcott, are very susceptible, while Persian limes (*C. aurantifolia*), true lemons (*C. limon*), sour limes (*C. aurantifolia*), sweet limes (*C. limettioides*) are practically resistant to the disease (BITANCOURT, 1955; ROESSING & SALIBE, 1967; RODRIGUES, 1995; BASSANEZI et al.., 2002; BASTIANEL et al., 2004; BASTIANEL et al., 2006).

The symptoms of leprosis are yellowish spots on the leaves, cortical lesions on the branches and round necrotic spots on the fruit. Severely affected plants show defoliation, premature and intense fruit drop, excessive drying of branches and the death of spikes, which can lead to the plant's death and reduce the longevity of the orchard (FEICHTENBERGER et al., 1997).

The disease is mainly controlled by applying chemical products to control the mite vector, which is why many studies have been carried out (NAKANO et al., 1987; OLIVEIRA et al., 1991; ANDRADE et al., 2010). However, some work has also been carried out with an emphasis on biological control with predatory mites and fungi (SILVA, 2005), as well as the search for plants resistant to the disease (BASTIANEL et al., 2006).

2.2 Bioecological aspects of the mite *Brevipalpus phoenicis*

Known in Brazil as the leprosis mite, or in English as the "apartment mites" or "false spider mite", the species *Brevipalpus phoenicis* belongs to the class Arachnida, subclass Acari, suborder Prostigmata, superfamily Tetranychoidea and family Tenuipalpidae. According to Albuquerque et al. (1997), the *B. phoenicis* mite was first described by Geijskes, in 1939, infesting *Phoenix* spp. plants in a greenhouse in Holland. However, due to its wide geographical distribution and the fact that it infests a variety of plants, it has not been possible so far to establish the place of origin of this mite, but it is believed to be in a region with a tropical climate.

The *Brevipalpus* genus includes more than 200 species (MESA et al., 2009) worldwide, and the *B. phoenicis* species is considered to be the most important for agriculture, as it is responsible for transmitting viruses to various plants (WELBOURN et al., 2003).

Kitajima et al. (2003) listed 25 viruses associated with *Brevipalpus* sp. in Brazil, including: *Citrus leprosis virus* (CiLV) in citrus, *Ligustrum ringspot virus* (LiRSV) in ligustrum, *Coffee ringspot virus* (CoRSV) in coffee, *Passion fruit green spot virus* (PFGSV) in passion *fruit*, *Hibiscus green spot virus* (HGSV) in hibiscus, *Malvaviscus ringspot virus* (MRSV) on mallow, *Ivy green spot virus* (IGSV) on ivy, *Clerodendron chlorotic spot virus* (CCSV) on clerodendron, *Orchid fleck virus* (OFV) on orchids and

Solanum Violaefolium ringspot virus (SVRSV) on solanum.

This species can be distinguished from other tenuipalpid mite species by its dorsal opisthosomal ketotaxis, with one pair of humeral setae and five pairs of dorsolateral setae. In addition to these characteristics, the females of the *B. phoenicis* species have two rod-shaped setae called solenids on tarsus II (WELBORN et al., 2003).

8. phoenicis is a polyphagous species and can be found on various host plants, including weeds commonly found in citrus groves (TRINDADE & CHIAVEGATO, 1994), so there is a possibility that these plants may serve as a refuge for the mite.

The mite is cosmopolitan, having been reported on all continents except the Arctic and Antarctic regions, and is polyphagous with approximately 456 hosts recorded, which proves its high degree of adaptation, making it difficult to control (CHILDERS et al, 2003a). In Brazil, the transmission of leprosis in field conditions is attributed exclusively to the mite *B. Phoenicis* which, when feeding on infected tissue cells, ingests viral viral particles, and subsequently infects healthy tissues when it feeds again (BASSANEZI, 2004).

The females have an oval body and are between 0.29 mm and 0.31 mm larger than the males. Both males and females are orange in color and darken with age, varying with food and climatic conditions; in the region of the digestive tract there is a darker area in the shape of the letter "H" on the idiosoma (HARAMOTO, 1969).

The biological cycle of the *B. phoenicis* mite consists of four active stages (larva, protonymph, deutonymph and adult) and four immobile stages (egg, protocrystallid, deutocrystallid and telocrystallid). The immobile stages are physiologically active and occur between the active stages (RODRIGUES et al., 2003). The complete cycle on citrus fruit is 14.4 days and 17.6 days when reared on citrus leaves at a temperature of 30° C (CHIAVEGATO, 1986).

Chiavegato (1995) proved that the mite, at any stage of its development, once infected with the virus, becomes a transmitter of the disease throughout its life; however, transovarian transmission does not occur. Although potentially all stages of the life cycle are equally transmissible, the adult stage is particularly important, given its greater mobility and longevity, which greatly increases the chances of becoming infected and spreading the disease.

The female lays eggs in protected places, mainly on citrus fruit with wart lesions (ALBUQUERQUE et al., 1995; ALBUQUERQUE et al., 1997). The egg is elliptical, measures around 0.100 to 0.116 mm in length and 0.06 to 0.08 mm in width and is bright orange in color. The larva has three pairs of legs, is orange in color and measures 0.13 x 0.08 mm. The protonymph is slightly larger than the larva (0.25 x 0.15 mm) and differs from the larva mainly in having four pairs of legs. After a new quiescent period, it transforms into the deutonymph (0.32 x 0.17 mm), which differs from the previous stage in that it is larger. The deutonymph undergoes a further period of dormancy to reach the adult stage. Sexual dimorphism is pronounced and the male has a tapered opisthosoma, with two transverse sutures dorsally, which differentiate it from the female, which has only one (OLIVEIRA & PATTARO, 2005). According to Haramoto (1969), the sex ratio is only 1 to 1.5% males.

The low occurrence of males in *B. phoenicis* populations and the mechanism of sex determination are unclear (PIJNAKER et al., 1981), as is the feminizing action caused by an endosymbiont *bacterium* belonging to the genus *Cardinium* (WEEKS et al. 2003; KITAJIMA et al. 2007). The reproduction of *B. phoenicis* occurs predominantly by telite parthenogenesis, producing only females identical to the progenitor.

B. phoenicis mites are haploid with only two non-homologous chromosomes (n=2). This is the only case described in the literature in which an organism never produces the diploid phase, nor when sexual reproduction occurs, which is rare, and takes place with the participation of males (HELLE et al., 1980; PIJNACKER et al., 1980; 1981; WEEKS et al., 2001).

2.3 The leprosis virus CiLV

The first considerations about the etiologic agent of citrus leprosis emerged with the first reports of the disease by Fawcett (1911), who attributed it to the action of fungi. However, Vergani (1945) proposed that the disease was caused by a toxin expelled by the mite's saliva during feeding. This hypothesis was maintained for several years, until the first evidence of the viral etiology of leprosis was found through its transmission by grafting (KNORR, 1968).

The first particles found were bacilliform, non-enveloped, rhabdovirus-like, measuring 100-110 nm long and 4050 nm wide, associated with viroplasms in the cell nucleus (KITAJIMA et al., 1972). Bacilliform particles measuring 120-130 nm long and 50-55 nm wide and associated with the endoplasmic reticulum lumen in the cytoplasm of parenchyma cells were also reported by Colariccio et al. (1995).

Nowadays, the leprosis virus has been associated with two distinct morphological types: the cytoplasmic type, which is more common (KITAJIMA et al., 1974; COLARICCIO et al., 1995), and the nuclear type (KITAJIMA et al., 1972). It is assumed that the two types of particles are different stages of development of the same virus or two different viruses (KITAJIMA et al., 1995).

However, in Brazil, the prevalent type of viral particle found is cytoplasmic (CiLV-C) (KITAJIMA et al., 2004). Salva & Massari (1995) found that more than 60% of orchards in the state of Sao Paulo showed symptoms of leprosis.

Pascon et al. (2006), after complete nucleotide sequencing of CiLV-C, cytoplasmic type, suggested that the virus is a probable member of the Tobamoviruses. However, Locali-Fabris et al. (2006) suggest that CiLV-C should be considered a type-member of a new genus of phytovirus, called *Cilevirus*.

CiLV particles are only observed in tissues damaged by leprosis and are not found in adjacent asymptomatic areas, which apparently indicates the non-systemic nature of the virus (KITAJIMA et al., 1972; COLARICCIO et al., 1995). According to Rodrigues et al. (2004) this non-systemicity of the virus considerably increases the importance of the vector in the epidemiology of the disease, since its occurrence is an essential condition for intra- and inter-plant dissemination in orchards.

2.4 Spirodiclofen and Cyhexatin

The acaricide spirodiclofen was recently registered for the control of *B. phoenicis* in citrus, with indications that it inhibits lipid synthesis, and therefore has a different mechanism of action to other acaricides on the market (POLETTI & OMOTO, 2004).

According to Oliveira & Pattaro (2004a), in tests of the direct action of spirodiclofen on *B. phoenicis*, it was found that 100% mite mortality was achieved 7 days after application of the product. Residual efficiency, on the other hand, decreases as the interval between application and transfer of the mites increases, and increases with the time the mites remain on the spirodiclofen residue.

In addition to its effects on adult mites, Spirodiclofen decreased the viability of *B. phoenicis* eggs.

As it has a different mechanism of action, spirodiclofen may not be cross-resistant with other acaricides on the market (FISCHER & BENET-BUCHHOLZ, 2002). However, Poletti & Omoto (2004) found that spirodiclofen does not present cross-resistance with the acaricides dicofol, propargite and hexythiazox.

With regard to the selectivity of spirodiclofen to predatory mites, Reis et al. (2005) evaluated the mortality and reproduction of adult females of predatory mites from the Phytoseiidae family, using residual action bioassays, and concluded that the acaricide spirodiclofen shows physiological selectivity to predatory mites.

Among other positive aspects, spirodiclofen is very stable in the environment (WACHENDORFF et al., 2002), and shows no tendency to volatilize in the atmosphere, being rapidly degraded and mineralized in the soil (BABCZINSKI, 2002).

The acaricide cyhexatin belongs to the chemical group of organotins, which act on mites by inhibiting oxidative phosphorylation, differing from the mode of action of other chemical groups of acaricides used on citrus (CORBETT et al., 1984). Therefore, the rotation of organotin acaricides with other groups is highly recommended (KONNO et al., 2001).

Despite the intense use of organotin acaricides, until 1999, no cases of resistance at levels above the critical frequency (> 10% of resistant individuals) were detected in *B. phoenicis* mite populations in citrus groves in the state of Sao Paulo (KONNO et al.

2001). However, it should be noted that cross-resistance has been reported between the organotin acaricides cyhexatin, fenbutatin oxide and azocyclotin in *Tetranychus urticae* (Koch, 1836) in Australia, and between cyhexatin and fenbutatin oxide in *T. urticae* and *Tetranychus pacificus* (McGregor, 1919) in the United States (FRANCO et al., 2007).

Fernandes et al. (2008), evaluating the efficiency of various acaricides in the laboratory *on B. phoenicis*, found cyhexatin to be highly efficient. The authors observed that *B. phoenicis* mite populations from coffee plantations and citrus orchards showed similar mortality under the action of cyhexatin.

Trials conducted by Reis et al. (1998) on coffee trees showed that cyhexatin was highly effective in controlling the coffee leaf spot mite *B. phoenicis.* Andrade et al. (2008) found high mortality of the mite *Tetranychus mexicanus* (McGregor, 1950) with the use of cyhexatin, both by direct and residual action.

2.5 Sulphocalcic syrup

Sulphocalcic syrup is considered a good product for so-called agroecological systems (PLANETA ORGÂNICO, 2003) to control pests and diseases. As well as being practically non-toxic to mammals, it is one of the only chemical products accepted by the IBD - Biodynamic Institute for Rural Development, the first national organic certifier recognized by IFOAM - International Federation of Organic Agriculture Movements in 1995 (HOFFMANN, 2004).

According to Pattaro & Oliveira (2005) sulphocalcic syrup is used to control mites, insects, mosses and lichens in various crops, but it is in citrus that it is most frequently used, mainly to control the mite *B. phoenicis* and the citrus rust *P. oleivora.*

The use of sulphocalcic syrup as an alternative product for controlling the mite *B. phoenicis* has been intensified by Brazilian citrus growers, especially since the development of organic citrus farming in recent years.

Sulphocalcic grout is an inorganic sulphate whose active ingredient is calcium polysulphide, obtained by dissolving powdered sulphur and virgin or hydrated lime.

According to Abbott (1945), the toxicity of the mixture is due to the release of the toxic gases hydrogen sulphide (H_2S) and sulphur dioxide (SO_2). Sulphocalcic syrup has a fungicidal, insecticidal and acaricidal action and is also used as a foliar fertilizer (PRATES, 1999). In addition, according to Bertoldo (2003), applications of sulphocalcic syrup can supply calcium and sulphur to plant metabolism, which stimulates photosynthesis reactions and induces greater resistance to pests.

The parameter used by many manufacturers to assess the quality of sulphocalcic grout a few years ago was the apparent density, measured by an aerometer on the degrees Bé scale, with the quality considered acceptable if the apparent density was between 27 and 30 degrees Bé. Nowadays, acceptable quality is considered to be that which has 15% polysulphides and high purity. It is important to note that sulphocalcic syrup is not protected by a patent and can be made by the farmer himself (PATTARO & OLIVEIRA, 2005) with a view to reducing costs.

Andrade et al. (2007) obtained, under laboratory conditions, good results for the efficacy of sulphocalcium solution on the Mexican spider mite *T. mexicanus*, by direct, residual and ovicidal action. Oliveira et al. (2002) found a high residual action of sulphocalcium solution on the rust mite *P. oleivora*. Oliveira et al. (2008) evaluated the efficiency of sulphocalcium mixture in coffee cultivation on the mite *Olygonychus ilicis*, and found that this product was highly effective in killing the mite under field conditions.

Pattaro (2003) found that sulphocalcium mixture, at dosages of up to 80 L/2000L of water, has low residual efficiency on adults of *B. phoenicis*. However, the efficiency increases as the period of exposure of the mite to the residue increases, although it still falls short of what would be desired for its control. On the other hand, the direct action is satisfactory, even at low concentrations of the product.

The efficiency of sulphocalcium solution on *B. phoenicis* eggs is low, around 36.4%, at a dosage of 80 L/2000 L of water. Once the larvae have hatched, their contact with the mixture's residue results in high mortality rates (PATTARO, 2003).

Due to the high efficiency of sulphocalcium solution through direct action on the adult

B. phoenicis mite, as well as its low cost, it has become common practice among producers to mix it with half a dose of an ovicidal product, known as a "tank mix" (PATTARO & OLIVEIRA, 2005). However, successful mite control with mixtures of products requires a low frequency of resistance, absence of cross-resistance and similar biological persistence for the two compounds. Pattaro (2003) also pointed out that successful control of *B. phoenicis* with sulphocalcium solution requires good coverage of the plants, using the right dosage, volume and application equipment, thus increasing the chances of directly hitting the mite.

According to Pattaro (2003), the control of leprosis and rust mites exclusively with sulphocalcic syrup has been shown to be economically viable compared to other acaricides, as long as the number of applications is not excessive. However, Pattaro & Oliveira (2005) do not recommend controlling these pests only with sulphocalcic syrup in successive applications, given the possibility of selecting resistant individuals.

According to Prates (1999), phosphorus insecticides should not be applied on the residue of sulphocalcium syrup, as they may lose their action due to the high pH on the surface of the leaves as a result of the calcium carbonate deposited on them. Pattaro (2003) observed that the residual action of the acaricides analyzed was not negatively influenced by the calcium carbonate left on the fruit after the spray. In the few acaricides where interference was found, it proved to be positive, with the residue of the sulphocalcium solution constituting an additive factor to the mortality of the *B. phoenicis* mite.

Some negative aspects related to the use of sulphocalcic syrup have been pointed out, such as phytotoxicity to fruit and shoots, sometimes caused by a single application at dosages of more than 80 L/2000 L of water, especially to those fruits most exposed to sunlight (PATTARO, 2003).

The mechanism of action of sulphocalcium syrup has not yet been fully elucidated, however, as its main active ingredient is sulphur, it is believed that the syrup acts as an inhibitor of cellular respiration (IRAC, 2010). Therefore, as there is no other product with a different mechanism of action accepted in organic citrus growing, its recommendation should always be careful and associated with other management strategies in order to hinder the selection of resistant mite populations (CASARIN, 2010).

2.6 Management of citrus leprosis

The main leprosy control measure is based on the application of acaricides to control the mite vector *B. phoenicis* (OMOTO, 1995; BASSANEZI, 2001). The absence of vector control, associated with periods of prolonged drought, leads to an increase in the mite's population, making it very difficult to control, which can lead to outbreaks of the disease, as happened in several years from 1981 to 2000 (BASSANEZI, 2001).

However, with regard to vector mite control, other measures should be adopted, such as: planting mite-free seedlings, disinfecting vehicles and collection boxes (ROSSETTI et al., 1997); eliminating mite-host weeds (TRINDADE & CHIAVEGATO, 1994; CATI, 1997; MAIA & OLIVEIRA, 2004); using practices that favor the population of natural enemies (YAMAMOTO et al., 1992); control wart infestation (CATI, 1997); avoid using brushcutters or harrows during the dry season (OLIVEIRA & PATTARO, 2004b); use green cover with species less favorable to mites (GRAVENA et al., 1992); use windbreaks and hedges that are unfavorable to the mite (FEICHTENBERGER, 2000), carry out chemical control of the mite using selective products with different mechanisms of action (POLETTI & OMOTO, 2004). To reduce sources of virus inoculum, it is recommended to eliminate weeds, hedges and windbreaks that are hosts to the virus (MAIA & OLIVEIRA, 2005; NUNES, 2004); plant healthy seedlings (OLIVEIRA, 2004); collect fruit with symptoms and fallen to the ground after harvest (BUSOLI, 1995); bring forward the harvest (OLIVEIRA, 1986) and prune parts attacked by the disease (PATTARO, 2006).

2.7 Physico-chemical characteristics of citrus fruits

Pozzan & Triboni (2005) reported that weight, fruit size, juice content, soluble solids, acids and vitamin C, juice color, and their derived indices such as *ratio* and industrial yield (total number of boxes needed to produce one ton of concentrated juice at 66°

Brix) are the most commonly used measures for characterizing the ripeness of fruit in a citrus grove.

Soluble solids (SS) are made up of all the constituents of citrus fruit that are dissolved in the water portion of the juice (TING, 1983), with soluble sugars and organic acids standing out in citrus fruit (SINCLAIR, 1960; ERICKSON, 1968). In fruit quality studies, the determination of the °Brix of juices is used instead of sugar concentrations (TING & ROUSEFF, 1986). They are measured in manual refractometers and require 2 to 3 mL of juice to take the readings, which range from 0 to 70 °Brix (SOUZA, 2009).

The acidity of citrus juice increases at the beginning of fruit development and remains practically constant until the initial stage of ripeness. The acidity decreases as the fruit matures, due to dilution by the increase in fruit size. The basic method for calculating acidity is based on titrating a known quantity of juice using sodium hydroxide (NaOH) and phenolphthalein as an indicator. The result is expressed as a % of citric acid. In terms of ripeness, acidity differs from soluble solids, as there is a large variation in total acidity and a smaller variation in soluble solids (AGUSTi et al., 1994; SOUZA, 2009).

There is an empirical relationship that is given by dividing the soluble solids content by the percentage of titratable acidity, called the *ratio*. The ratio 13, for example, means 13 parts soluble solids for each part acid. The lower the ratio, the more acidic the juice, and vice versa (SOUZA, 2009). The *ratio* can be used as a ripeness test, because the soluble solids increase and the acids decrease during the growth and ripening of the fruit (BARTHOLOMEW

& SINCLAIR, 1943). Although the ratio of soluble solids to acidity only describes the taste of the fruit, it is the best ripeness index available, as it is easy to determine and is close to the actual degree of ripeness (TING, 1983).

The industrial yield is given by the technological index, which is represented by the physical and chemical characteristics of the fruit. As well as being an indicator of maturity, the technological index can also be used as an indicator of fruit quality (SOUZA, 2009; SOUZA & GOES, 2010).

The variables used to determine the ripeness of citrus fruit are dynamic throughout the harvest. In general terms, the *ratio* evolves linearly, increasing on average until November for the Pera variety, varying by as much as 300% compared to April. The accumulation of soluble solids shows a quadratic evolution model, with a peak in accumulation in the spring months, while the vitamin C content of the juice increases, reaching 80% of the April value at the end of the ripening cycle. The mass of the fruit varies mainly according to the availability of water in the soil from May onwards, so the fruit normally loses mass in the winter months, in many cases withering sharply, and with the return of water availability, from the spring rains onwards, there is a rapid and continuous gain in mass or size until it is harvested (POZZAN & TRIBONI, 2005).

Citrus fruits are non-climacteric and therefore do not have a clear point indicating ripeness, since ripening is characterized by a gradual increase in juice, a decrease in acidity and an increase in the amount of soluble solids and the *ratio* (AGUSTi et al., 1994; SOUZA, 2009). The behavior of fruit ripening characteristics throughout the season is basically a function of the temperature and the specific water regime of each area, which explains some of the variations that occur in the ripening period of fruit when comparing different seasons (VOLPE, 2002).

According to Pozzan & Triboni (2005), citrus juices with a *ratio* between 14 and 16 are the most appreciated by consumers all over the world, due to the balance, in sensory terms, between the sugar and acid content. As far as citrus varieties are concerned, 'Pera' stands out because it has the best colored juice of all the processed cultivars and also has a high juice yield, often reaching over 60% (POZZAN & TRIBONI, 2005).

The 'Pera' orange is the most widely grown in the state of Sao Paulo and certainly the most important for the national citrus industry, due to the characteristics mentioned above, as well as the fact that the fruit is oval-shaped, with three to four seeds and an average size of between 146 and 156 g/fruit. It has a thin rind with a good number of oil vesicles, the average oil content of which ranges from 0.6 to 0.8% by weight. It provides a good industrial yield, with an average soluble solids content of over 11.5%. It is considered to be a mid-season cultivar, with the main harvest taking place between

July and September.

Rodrigues (2000) assessed the quantitative and qualitative damage caused by leprosis to the fruit of various citrus varieties. The author found that as the number of lesions on the fruit increased, there was a reduction in fruit mass. This was also found by Chiavegato & Salibe (1981), who correlated the number of lesions on fruit affected by leprosis with their mass, and found that there was a significant reduction in fruit mass as the number of lesions increased.

Rodrigues (2000) stated that the reduction in fruit mass is apparently due to a significant reduction in fruit width, as the fruit shows a greater number of lesions. The regression between fruit height/width is therefore significant, and the number of fruits needed to complete a 40.8 kg box was directly proportional to the number of leprosy lesions quantified on the fruit. However, Rodrigues (2000) found that the variables related to the fruit's chemical characteristics were not affected by citrus leprosis lesions.

3, METHODOLOGY

3.1 Experiment site

The experiment had been installed since October 2003 at Fazenda Sao Pedro (coordinates: latitude 49° 13' 24" W, longitude 21° 56' 26" S and altitude 468 m), a property belonging to the Branco Peres Group, located in the municipality of Reginópolis, SP.

The soil in the experimental area was classified as Latossolo Vermelho- Amarelo (LVA), and the plants used were sweet orange *Citrus sinensis* (L.) Osbeck cv. 'Pera', grafted onto Cleopatra mandarin (*Citrus reshni* Hort), 12 years old when the experiment was set up, spaced 7 x 4 meters and irrigated by drip irrigation.

3.2 Setting up the experiment

Initially, to set up the experiment, approximately 500 citrus plants were inspected to determine the level of infestation of the *B. phoenicis* mite, using the conventional sampling method (CATI, 1997) and it was found that the level of infestation was below 5%.

The experimental design was randomized blocks, with three treatments and 16 replications. The treatments used were: (1) Spirodiclofen and cyhexatin applied in rotation; (2) sulphocalcic syrup; (3) no acaricide applied. Each experimental plot consisted of 3 plants arranged in a row. The acaricide applications were based on the level of infestation of the *B. phoenicis* mite. Approximately seven (7) years after the start of the experiment, the differences in the severity of leprosis between these treatments are clearly evident (Figure 1). Spirodiclofen and cyhexatin were used at a dosage of 20 and 50 mL p.c./100 L of water, respectively, and sulphocalcic syrup at a dosage of 4,000 mL p.c./100 L of water and 22° bé.

Figure 1. Visual aspect of citrus plants treated and not treated with acaricides: (A) no acaricide; (B) sulphocalcic syrup and (C) spirodiclofen/cyhexatin (May 2010).

3.3 Determining the level of infestation of the mite *Brevipalpus phoenicis*

Population surveys of the mite *B. phoencis* were carried out every fortnight after the experiment was set up. In each plot, the central plant was sampled by examining three fruits located inside the plant canopy and with the presence of wart disease, using a 10x magnification field magnifier (MARTINELLI et al., 1976). In the absence of fruit, three branches, at the beginning of suberification, approximately 25 cm long, were evaluated. To determine the level of infestation, fruit with at least one mite at any stage of development was considered to be infested. The products were applied based on the level of *B. phoenicis* infestation, adopting a control level of 8.3% (PATTARO, 2003).

The products were applied using a tractor-drawn sprayer with hand-held lances, using a volume of spray of 15 to 16 L/plant to provide complete coverage of the plants. The other phytosanitary treatments, with the exception of *B. phoenicis* mite control, were carried out as normal, giving preference to products that are more selective to mites.

3.4 Assessments of leprosis severity in citrus plants

Assessments of leprosis severity on all the plants in the experiment were carried out in July 2009, July and October 2010, based on a visual grading scale (Table 1).

Table 1. Visual grading scale (0-5) for assessing the severity of leprosis in citrus plants (RODRIGUES, 2002).

Notes	Symptoms of leprosis in citrus plants
0	No injuries
1	Few lesions in any organ, restricted to a sector of the plant
2	Lesions in more than one organ and/or distributed in more than one sector
3	Lesions abundant in all organs and well distributed throughout the plant
4	Abundant lesions (whole plant) and leaf and/or fruit drop
5	Previous (4) + drought and death of branches.

3.5 Technological analysis of citrus fruits

Ten fruits were randomly collected from the central plant of each plot, about 3 days before the actual harvest. The samples were collected in August 2009, August and December 2010. The samples were sent to the Chemical Analysis Laboratory at the Bebedouro Citriculture Experimental Station - EECB, Bebedouro - SP, to determine the fruit's physical and chemical characteristics. The characteristics determined were:

a) Average fruit mass: determined by gravimetry using a 12 kg digital scale. The average mass was obtained from the ratio between the mass of all the fruit in each treatment and the number of fruit.

b) Average fruit diameter: the transverse and longitudinal diameters of each fruit were measured using a graduated caliper (cm).

c) Juice yield: the juice from the fruit was extracted using a Centenàrio Model OTTO 1800 juice extractor. The extracted juice was then sieved to determine the volume, expressed in mL/100 g of fruit, using a graduated beaker. The juice yield was expressed as a percentage.

d) Soluble solids (SS): determined by refractometry at 20° C and expressed in °Brix (g/100 mL) according to methods established by Redd et al. (1986).

e) Titratable acidity (TA): determined by titration with 0.1 N NaOH solution (normal), using phenolphthalein as an indicator and a Tecnal digital bench pH meter. Based on the amount of NaOH consumed, the TA was determined. The results were expressed in grams of citric acid per 100 mL of juice.

f) ripeness index or *ratio*: calculated as the ratio between SS content and TA.

g) Technological index (TI): refers to the amount of SS per 40.8 kg citrus box. It was calculated using the equation proposed by Di Giorgi et al. (1990) cited by Silva (1999).

$$IT = SST * RS * 40.8 / 10000$$

where: IT = technological index, SS = soluble solids (°Brix), RS = juice yield and 40.8 is the industrial standard weight in kg of the box of oranges.

3.6 Determining climate variables

Rainfall, temperature and relative humidity were recorded daily during the 2008-2009 and 2010-2011 harvests, when the fruit was collected for analysis.

3.7 Productivity assessment

The productivity of the treatments was assessed by harvesting all the fruit from each plot. The harvested fruit was then weighed using a Plenna® dynamometer scale with a maximum capacity of 50 kg. Two harvests were quantified, in August 2009 and August 2010.

3.8 Form of data analysis

The data obtained was submitted to analysis of variance using the F test and the means were compared using the Tukey test at 5% probability. Before being submitted to analysis of variance, the data was transformed into ln (x +5) in order to normalize the data. The values of the fruit's physico-chemical characteristics were correlated with the average values of leprosis severity obtained in each experimental plot. The variation in the fruit's physicochemical variables between samples was also analyzed.

4. RESULTS AND DISCUSSION

4.1 Climatic data during the experiment period

The physical and chemical characteristics of the fruit are greatly influenced by climatic factors, especially those related to rainfall and temperature. It should be noted that 2009 had the highest accumulated annual rainfall, with a total of 1,241 mm (Figure 2).

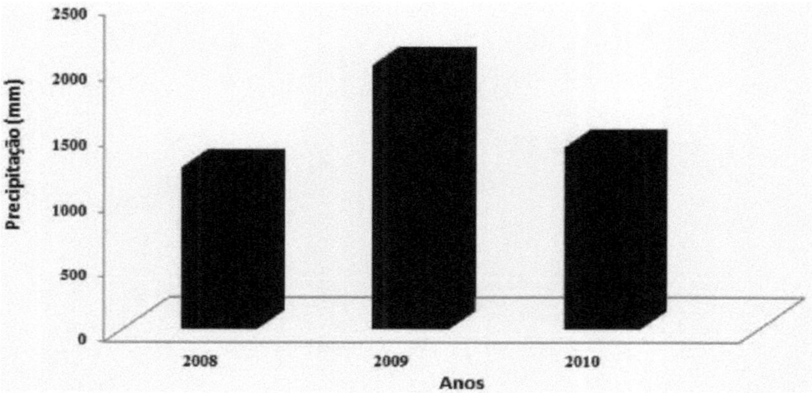

Figure 2: Total annual rainfall (mm) measured in the experimental area. Reginópolis - SP

According to Bartholomew & Sinclair (1943), climatic factors have the greatest influence on the development of citrus fruit.

The amount of water in the soil directly influences some components of the juice, for example, the soluble solids content can decrease with increased water availability, a fact also observed in the case of acidity (CRUSE et al., 1982). This fact was confirmed at the end of the 2008-2009 harvest, because due to the higher rainfall in 2009, the fruit collected in August 2010 had lower acidity and lower soluble solids content on average.

Several authors, including Di Giorgi et al. (1991), Tubelis & Salibe (1998) and Camargo et al. (1999), have found that rainfall is one of the main meteorological variables influencing citrus yields, mainly during flowering and initial fruit growth. Cunha Sobrinho et al. (1992) considered temperature to be another important environmental factor, among other factors also influenced, for example, fruit size, shape, skin color and ripeness.

Figures 3, 4 and 5 show the results for average monthly temperature and relative humidity (at Fazenda Sao Pedro - Reginópolis - SP) in the years in which the fruit was collected for analysis. It can be seen that there were no marked differences in these climatic factors between the years. Citrus fruits in conditions of high humidity and high temperatures have high juice and sugar production (BUSLIG, 1991). In this context, Albrigo (1990) showed that 60% to 70% of the variability between agricultural years for the technological index and total soluble solids in the juice may be due to variations in air temperatures and rainfall during flower induction and the differentiation period, before flowering.

However, recent studies have found that the pattern of viral infection caused by CiLV-C visualized in damaged tissues is little affected by climatic conditions, indicating that this virus has little genetic variability (BASTIANEL et al., 2010). Studies have shown that tissues damaged by CiLV-C collected in Central and South American countries showed similar patterns of viral infection to tissues collected in Brazil, even under different climatic conditions (MARQUES et al., 2010; CALEGARIO et al., 2011).

Generally speaking, leprosis is more severe in regions with a hot climate, as is the case in the north and northwest regions of the state of São Paulo, due to the prolonged periods of drought which encourage the vector mite to increase in population (BASSANEZI et al., 2002), and the high temperatures encourage CiLV-C to replicate in the plant's tissues (FREITAS-ASTUA et al., 2007).

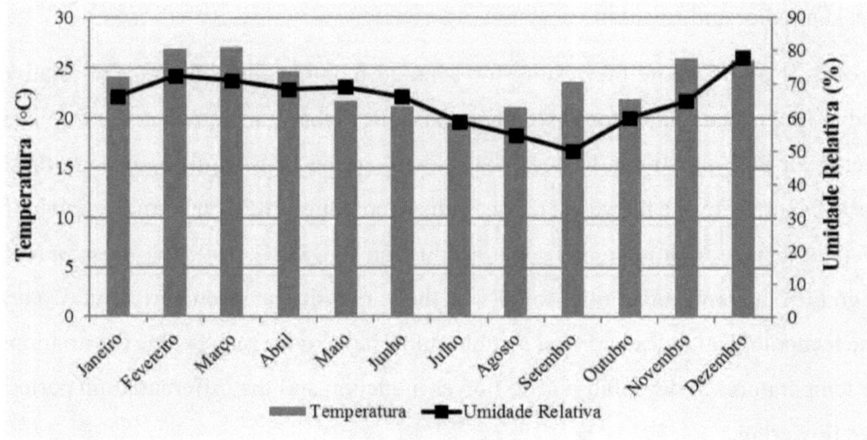

Figure 3: Monthly averages of temperature and relative humidity in 2008.

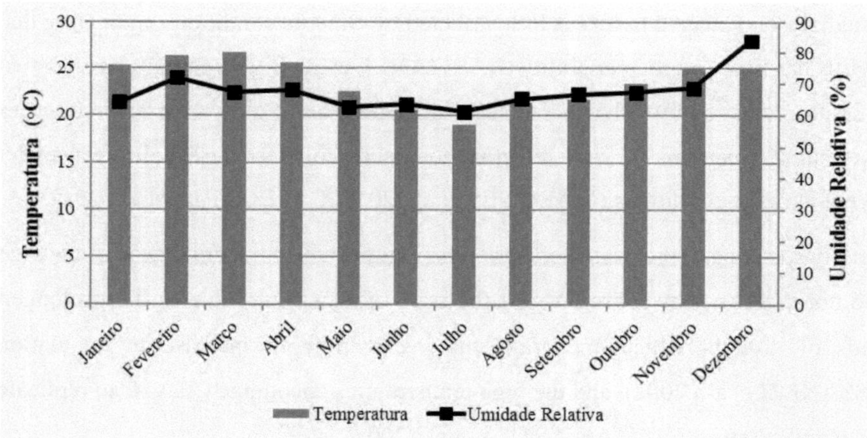

Figure 4 - Monthly averages of temperature and relative humidity in 2009.

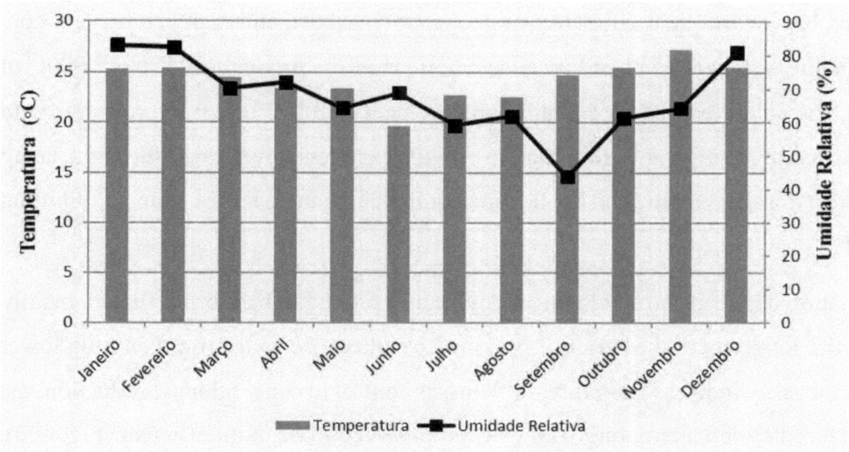

Figure 5 - Monthly averages of temperature and relative humidity in 2010.

4.2 *B. phoenicis* population fluctuation *and* acaricide applications

Figure 6 shows the results for the population fluctuation of the *B. phoenicis* mite, from the start of the experiment until March 2010, the month in which the acaricide applications were completed. Since the population of the *B. phoenicis* mite in the experimental area was below that established for its control.

Applications of the acaricides began in March 2005, when the *B. phoenicis* mite population reached control level (Figure 6). The rotation between the acaricides spirodiclofen and cyhexatin was effectively started in 2006; in 2005 only spirodiclofen was used. This rotation was adopted in order to reduce selection pressure, as the two products belong to different chemical groups and have different mechanisms of action.

A total of 31 applications of sulphocalcic syrup were made, 8 of spirodiclofen and 6 of cyhexatin. The greater number of applications of sulphocaecal syrup was mainly due to the low residual and ovicidal efficiency of this product (PATTARO, 2003). In addition, the continuous use of sulphocaecic syrup may have increased the frequency of resistant individuals, due to the selection pressure exerted, compromising its efficacy (CASARIN, 2010).

Pattaro (2003) found that sulphocalcium mixture, at dosages of up to 80 L p.c./2,000L of water, has low residual efficiency on *B. phoenicis* adults, although, by direct action, it was highly efficient, even at low concentrations of the product (20 L p.c./2,000L of water). The author emphasized that, in order to successfully control *B. phoenicis* with sulphocalcium solution, it is important to provide excellent coverage of the plant, using the correct dosage, volume and application equipment to increase the chances of hitting the mite.

The rotation between spirodiclofen and cyhexatin proved to be more efficient, mainly due to the longer period of control provided by the *B. phoenicis* mite and the lower leprosis severity indices. Therefore, for leprosis control in conventional production, the use of spirodiclofen alternating with cyhexatin proved to be more efficient (Figure 6). In general, the longest period of control of the mite *B. phoenicis* was obtained with the use of spirodiclofen.

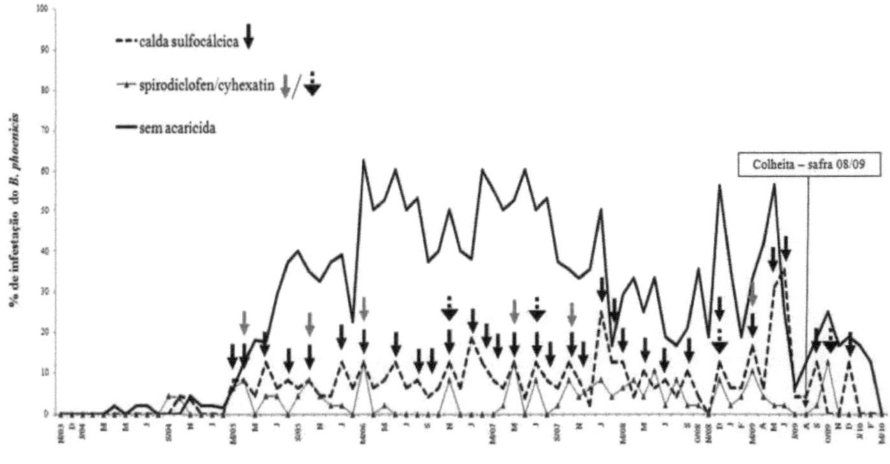

Figure 6 - Population fluctuation of the mite *Brevipalpus phoenicis* on 'Pera' orange plants, and the respective applications of acaricides (from November 2003 to March 2010). Reginópolis-SP.

4.3 Severity of leprosis

In the evaluation of the severity of leprosis in the plants of the experiment carried out in October 2010, it can be seen that the severity of leprosis was lower in the treatments with spirodiclofen/cyhexatin and higher in the control.

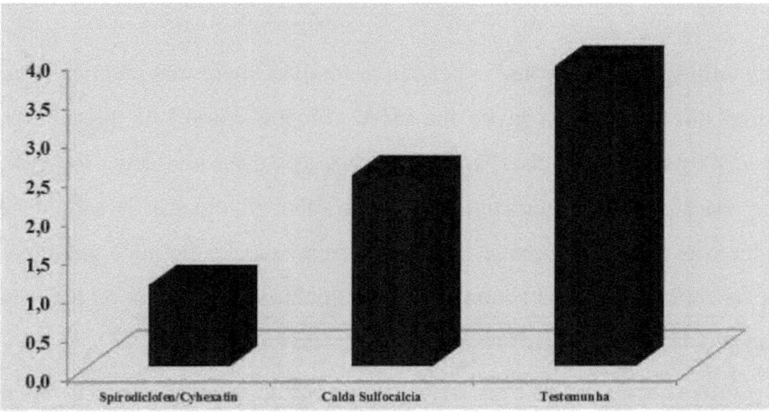

Figure 7. Severity of leprosis in plants treated or not with acaricides in the experiment to evaluate the physico-chemical characteristics of the fruit.

4.4 Leprosis x physical and chemical characteristics of fruit

4.4.1 2008-2009 harvest

The degree of severity showed no significant correlation with any of the variables analyzed for the fruit collected a few days before harvest (August 2009) (Figure 8). It is therefore assumed that the fruit that remained on the plants, with high and intermediate severity until harvest, were not reduced in size and mass due to leprosis, and are considered similar to the fruit from plants free of leprosis or with low severity. Juice yield, expressed as a percentage, also showed no significant correlation with leprosis severity.

The average fruit weight of the plants treated with sulphocalcium syrup was higher than that of the other treatments, with significant differences for the treatment with spirodiclofen and cyhexatin in rotation (Table 2).

In contrast to the treatment without acaricide, it was observed that the plants treated with sulphocalcic syrup had less leaf fall and branch death. Thus, it can be assumed that the greater leaf area maintained on the plants treated with sulphocalcic syrup led to greater absorption of water and nutrients by the roots, as well as a higher photosynthetic rate, which resulted in a greater mass of fruit that remained until harvest.

In the treatment with spirodiclofen and cyhexatin in rotation, there was practically no fruit or leaf drop due to leprosis, given the more efficient control of the mite *B. Phoenicis provided* by these products. *Phoenicis* provided by these products. However, the fruit mass differed significantly from the other treatments. It cannot be assumed that this was due to the greater quantity of fruit because the juice yield was not significant. Therefore, it can be assumed that the thickness of the skin on the fruit from the sulphocalcic syrup treatment resulted in a greater fruit mass.

Table 2. Physical characteristics of fruit collected in August 2009, as a function of different acaricides used to control the mite *Brevipalpus phoenicis*.

Acaricide	**Fruit diameter (cm)**		**Pasta**	**Income**
	transversal	**longitudinal**	**Fruit (g)**	**juice (%)**
No acaricide	6,5 a	7,2 a	181.1 ab	53,2 a
Sulphocalcic syrup	6,8 a	7,2 a	194,8 a	53,2 a
Spirod./Cyhex.	6,8 a	7,1 a	176,7 b	55,0 a
C.V.(%)	0,7	6,1	9,5	4,5

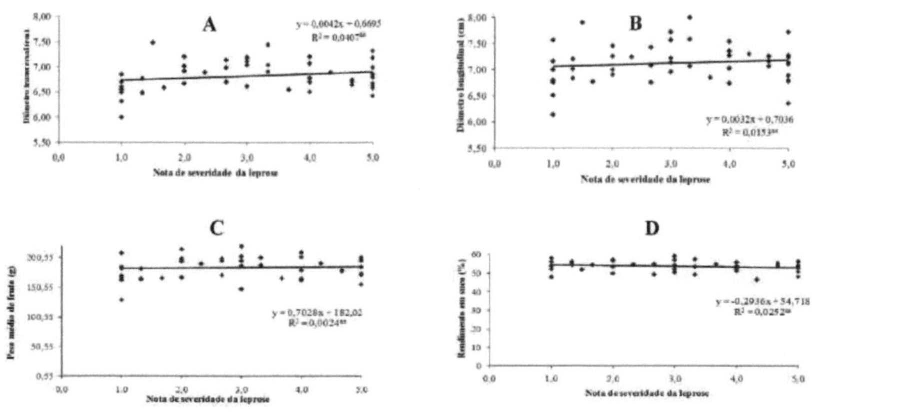

Figura 8. Correlations between fruit physical characteristics and leprosis severity, assessed on citrus plants (August 2009). (A) average transverse fruit diameter X leprosis severity; (B) average longitudinal fruit diameter X leprosis severity; (C) average fruit weight X leprosis severity and (D) average juice yield X leprosis

severity.

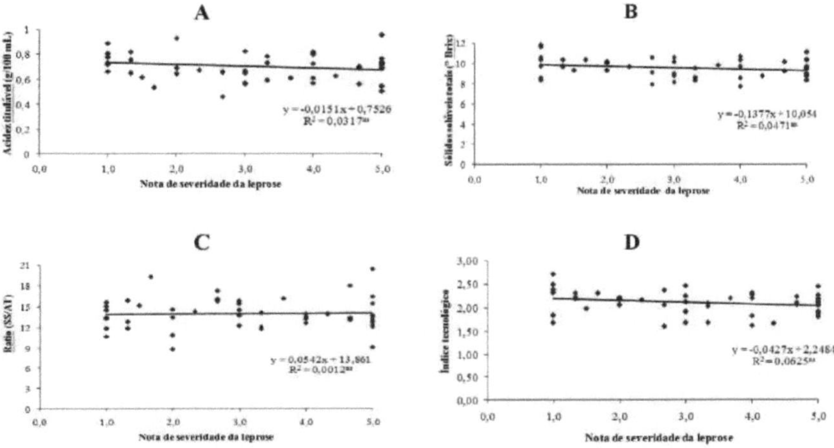

Figura 9. Correlations between fruit chemical characteristics and leprosis severity, assessed on citrus plants (August 2009). (A) titratable acidity X leprosis severity; (B) soluble solids X leprosis severity; (C) *ratio* X leprosis severity and (D) technological index X leprosis severity.

As for the chemical characteristics of the fruit analyzed: titratable acidity, °Brix and *ratio,* there was no significant correlation with the severity of leprosis. As for the industrial yield or technological index, which takes into account both the physical characteristics (juice yield) and the chemical characteristics (soluble solids) per 40.8 kg box of standard oranges, there was no significant correlation with leprosis severity.

The variables titratable acidity, °Brix and fruit *ratio* showed no statistical differences in relation to the acaricides used to manage citrus leprosis (Table 3). However, there were statistical differences in relation to the technological index. It was found that the treatment with sulphocalcic syrup obtained an average technological index value of 2.0, statistically different from the treatment with Spirodiclofen and cyhexatin in rotation, but not from the treatment without acaricide (Table 3). However, as no significant difference was found in subsequent harvests, this data was not discussed.

Table 3. Chemical characteristics of fruit collected in August 2009, as a function of different acaricides used to control the mite *Brevipalpus phoenicis*.

Acaricide	AT (g/100 mL)	SS (° Brix)	*Ratio* (SS/AT)	index Technology
No acaricide	0,7 a	9,5 a	14,0 a	2.1 ab
Sulphocalcic syrup	0,7 a	9,3 a	13,7 a	2,0 b
Spirod./Cyhex.	0,7 a	10,1 a	14,4 a	2,3 a
C.V.(%)	3,4	2,4	3,52	1,72

It is important to note that in the literature, studies verifying the effects of pests and diseases on these variables are very rare. Carrying out studies of this nature, which assess the influence of diseases and pests on technological indices, could help, in part, to plan the harvesting of orchards, with a view to improving the quality of the juice, optimizing soluble solids and the palatability of the fruit, as well as reducing productivity losses.

4.4.2 2009-2010 harvest

4.4.2.1 Fruits harvested in August 2010 (1ª flowering)

Similar to the results obtained in the previous harvest, the physico-chemical characteristics of the fruit were not significantly influenced by the severity of leprosis in plants treated or not with acaricides (Figures 10 and 11).

These results corroborate Rodrigues (2000) who found that leprosis does not affect the main characteristics of the juice when compared to juice from leprosis-free fruit.

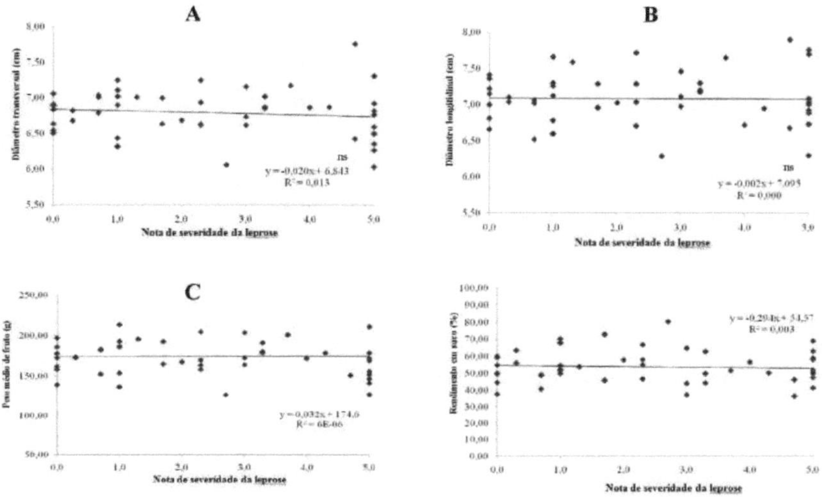

Figura 10. Correlations between fruit physical characteristics and leprosis severity, assessed on citrus plants (August 2010). (A) average transverse fruit diameter X leprosis severity; (B) average longitudinal fruit diameter X leprosis severity; (C) average fruit weight X leprosis severity and (D) average juice yield X leprosis severity.

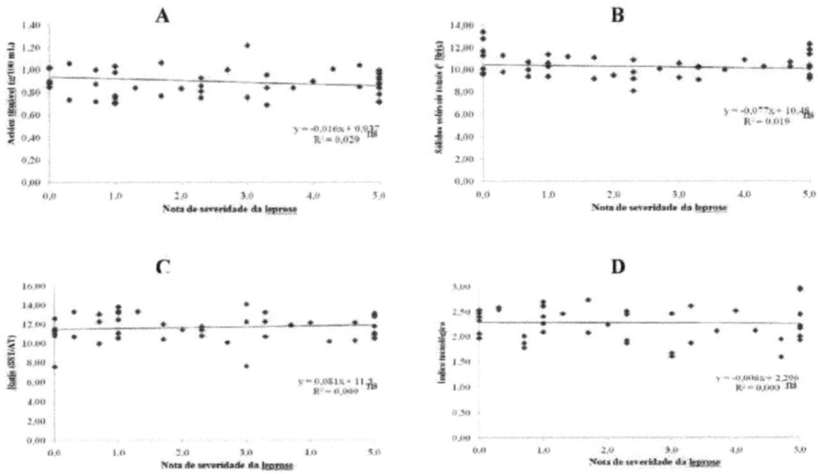

Figura 11. Correlations between fruit chemical characteristics and leprosis severity, assessed on citrus plants (August 2010). (A) titratable acidity X leprosis severity; (B) total soluble solids X leprosis severity; (C) *ratio* X leprosis severity and (D) technological index X leprosis severity.

Despite the great economic and social importance of leprosis in the Brazilian citrus agro-ecosystem, studies on the direct and indirect effects of this disease on fruit characteristics, especially those related to the specific chemical quality characteristics of the juice, are still scarce. This can be explained, in part, by the fact that leprosis is a disease that causes enormous damage to citriculture by drastically reducing productivity and the lifespan of plants, and so work has basically focused on controlling the mite-vector.

Several authors consider diseases to be a factor that can directly influence fruit ripening. Infections caused by diseases, especially those caused by viruses, due to their greater specificity with the plant, generally have significant effects on plant metabolism, affecting several essential cellular processes such as the synthesis of nucleic acids and proteins, lipids and the metabolism of carbohydrates and hormones, and enzyme functions (HULL, 2002). According to the same author, in plants severely attacked by viruses, respiration is usually lower than normal, but often no change occurs. However, in some plant-virus interactions, the rate of respiration increases even before symptoms appear, and continues to increase as the disease progresses.

According to Pascholati (1995), plants attacked by viruses do not normally suffer any change in transpiration rate, but he points out that symptoms that lead to a rupture of the cutinized surface of the plants cause an increase in transpiration rate. However, studies related to the effect of leprosis on the water balance of the host plant should be carried out separately, as it is apparently a non-systemic disease, unusual among plant diseases caused by viruses, which in the vast majority of cases are systemic.

It was found that the values obtained for the size and average mass of fruit from plants treated without acaricide did not differ statistically from the other treatments, with sulphocantbicide and with spirodiclofen and cyhexatin in rotation (Table 4). In these plants, with high leprosis severity, there was an intense fall of leprosis-injured fruit throughout the harvest, which suggests that the fruit that remained on the plants until close to harvest had a similar mass and size to the leprosis-controlled plants, in agreement with the results obtained in the previous harvest.

The results of this study showed that the fruit from plants free of *B. phoenicis* mite control *and* with high leprosis severity, which remained on the plant until harvest, had mass values identical to those of plants with mite control. According to Medina et al. (2005), the fruit is the most important drain on the plant's photoassimilates and, as a result of the intense fruit drop, there may be a higher rate of photosynthate translocation to the remaining fruit.

Table 4. Physical characteristics of fruit collected in August 2010, as a function of different acaricides used to control the mite *Brevipalpus phoenicis*.

Acaricide	Fruit diameter (cm)		Pasta	Juice yield (%)
	transversal	longitudinal	Fruit (g)	
No acaricide	6,8 a	7,0 a	169,0 a	54,7 a
Sulphocalcic syrup	6,7 a	7,0 a	168,4 a	56,0 a
Spirod./Cyhex.	6,9 a	7,2 a	177,9a	54,1 a
C.V.(%)	1,1	1,3	2,4	5,0

Table 5 shows the results of the analysis of the fruit's chemical characteristics as a function of the acaricides used. It can be seen that there were no differences in the variables titratable acidity, soluble solids, ratio and technological index. The highest *ratio* value was obtained in the treatment with spirodiclofen and cyhexatin in rotation, however, with no statistical difference from the other treatments.

Table 5. Chemical characteristics of fruit collected in August 2010, as a function of different acaricides used to control the mite *Brevipalpus phoenicis*.

Acaricide	AT	SS	*Ratio*	index
	(g/100 mL)	(° Brix)	(SS/AT)	Technology
No acaricide	0,8 a	10,3 a	12,2 a	2,3 a
Sulphocalcic syrup	0,9 a	9,9 a	11,4 a	2,3 a
Spirod./Cyhex.	0,9 a	10,5 a	11,3 a	2,3 a
C.V.(%)	1,6	2,5	3,1	3,2

3.2.1.2 Fruits harvested in December 2010 (2ª flowering)

In the fruit collected in December 2010, referring to the second flowering of the 2009-2010 harvest, the results obtained were similar to the previous results. The physical and chemical variables of the fruit evaluated showed no significant correlation with the severity of leprosis (Figures 12 and 13). Therefore, in the three fruit collections carried out, the physical parameters were not influenced by the severity of leprosis, proving that the fruit that remains on the plant until harvest is identical in physical terms, regardless of the intensity of leprosis.

With regard to viral plant diseases, Laranjeira et al. (2005) reported that they are the diseases that most alter the use of photosynthates compared to other plant pathogens. According to these authors, pathogens belonging to the virus group are highly specialized and their life cycle is closely linked to the host. As their survival depends on the host, they only eventually kill it, and in most cases they do not eliminate the plant's photosynthetic capacity, but only alter the redirection of photosynthate use.

The CiLV virus is non-systemic in the plant, unlike most plant disease-causing viruses that have the ability to translocate systemically in the plant. In this context, Freitas-Astùa et al. (2007), working with *C. sinensis* cv. Pera plants infected and not infected with CiLV, found, at the molecular level, a repression in the expression of genes related to photorespiration at the beginning of the infection of Pera oranges with CiLV, even before the appearance of visible lesions.

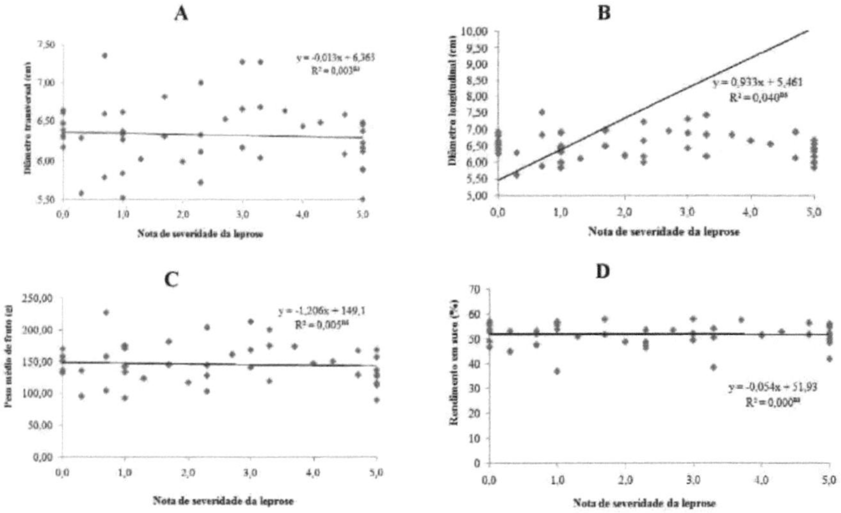

Figura 12. Correlations between fruit physical characteristics and leprosis severity, assessed on citrus plants (December 2010). (A) average transverse fruit diameter X leprosis severity; (B) average longitudinal fruit diameter X leprosis severity; (C) average fruit mass X leprosis severity and (D) average juice yield X leprosis severity.

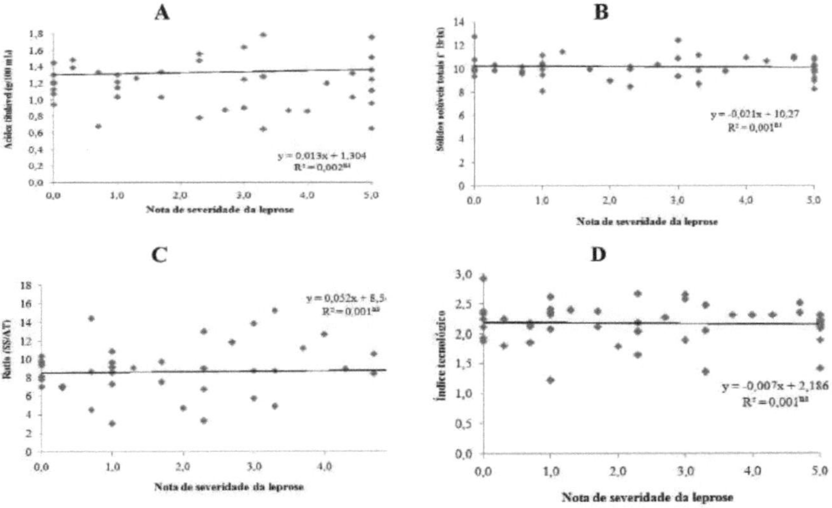

Figura 13. Correlations between fruit chemical characteristics and leprosis severity, assessed on citrus plants (December 2010). (A) titratable acidity X leprosis severity; (B) total soluble solids X leprosis severity; (C) *ratio* X leprosis severity and (D) technological index X leprosis severity.

Table 6. Physical characteristics of fruit collected in December 2010, as a function of different acaricides used to control the mite *Brevipalpus phoenicis*.

Acaricide	Fruit diameter (cm)		Pasta	Income
	transversal	longitudinal	Fruit (g)	juice (%)
No acaricide	6,2 a	6,4 a	132,6 a	50,0 a
Sulphocalcic syrup	6,4 a	6,7 a	150,3 a	51,0 a
Spirod./Cyhex.	6,4 a	6,5 a	147,6 a	53,7 a
C.V.(%)	1,4	1,2	3,2	2,1

Table 7. Chemical characteristics of fruit collected in December 2010, as a function of different acaricides used to control the mite *B. phoenicis*.

Acaricide	AT	SS	*Ratio*	index

	(g/100 mL)	(° Brix)	(SS/AT)	Technology
No acaricide	1,3 a	10,1 a	8,4 a	2,1 a
Sulphocalcic syrup	1,4 a	10,1 a	8,0 a	2,1 a
Spirod./Cyhex.	1,2 a	10,4 a	8,6 a	2,3 a
C.V.(%)	3,9	2,6	8,2	2,4

4.5 Citrus plant productivity

Figures 14 and 15 show the yield results, in orange boxes of 40.8 kg per hectare, for the harvests carried out in August 2009 (2008-2009 harvest) and August 2010 (20092010 harvest), respectively. In the 2008-2009 harvest, the treatment with spirodiclofen/cyhexatin showed the highest productivity, with an average of approximately 650 boxes per hectare, differing significantly from the treatment without acaricide application. On the other hand, in the subsequent 2009-2010 harvest, the average yield decreased in all three treatments and there were no significant differences between the treatments.

However, compared to commercial orchards, the average productivity was low in all the treatments. These figures are due to various parameters related to the crop, such as the age of the orchard, which in 2010 was 18 years old. In addition to leprosis, there were other diseases, especially black spot and CVC, and considerable reductions in fertilization, fertigation and liming promoted by the farm's management. During these years, the company's investments in this orchard were drastically reduced due to the sharp increase in production costs, mainly due to the costs needed to manage HLB (greening), in addition to the fact that the value of the orange box paid by the industry was very low.

As a result, there were considerable reductions in the use of phytosanitary products, fertilization and liming, as well as in the manpower available for inspections for other diseases and pests (black spot, CVC, mealybugs, etc.) and applications of phytosanitary products, which resulted in fruit losses and low plant productivity, partly explaining the low productivity obtained in these years of work. It is important to mention that citrus is a crop that has alternating production, which explains the decrease in productivity (GOLDSCHMIDT & GOLOMB, 1982; SOUZA, 1990).

In view of the results obtained so far, we can see that it is important to study the different aspects related to the leprosis pathosystem in greater depth, seeking greater efficiency in the control tactics used, providing more alternatives for monitoring, managing and perhaps eradicating the disease in orchards, as well as helping to reduce the damage caused to the citrus agro-ecosystem (RODRIGUES, 2000).

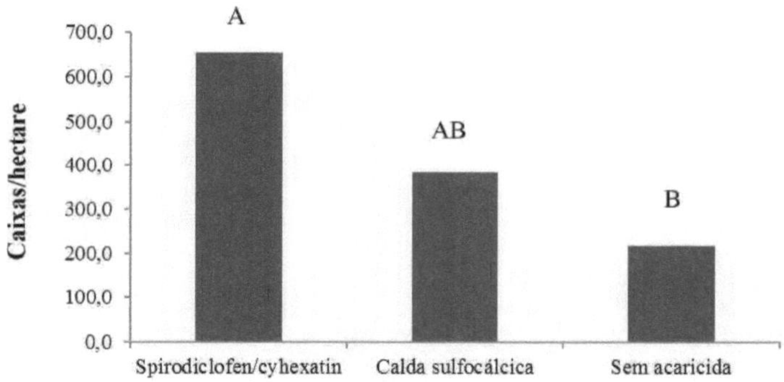

Figura 14. Average yield of citrus plants in cx. weight/hectare, as a function of different acaricides used to control the mite *Brevipalpus phoenicis*, referring to the harvest carried out in August 2009 (2008-2009 harvest). Reginopólis-SP, 2009.

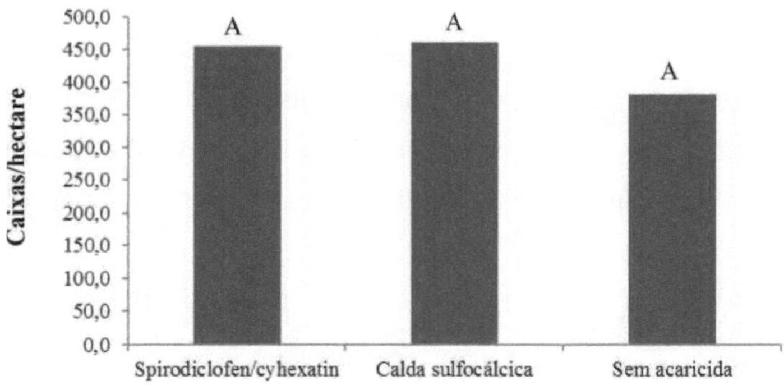

Figura 15. Average productivity of citrus plants in cx. weight/hectare, as a function of different acaricides used to control the mite *Brevipalpus phoenicis*, referring to the harvest carried out in August 2010 (2009-2010 harvest). Reginopólis-SP, 2010.

4.6 Variation in leprosy and physico-chemical characteristics between collections

Figure 16 clearly shows the differences in leprosis severity between the treatments. In the severity assessment carried out in July 2010 there was a considerable increase in leprosis in all three treatments, but in the assessment carried out in October 2010 there was a reduction in the average severity.

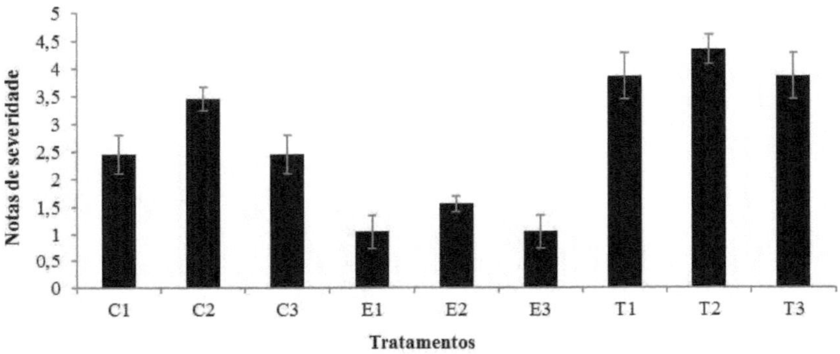

Figure 16. Average values of leprosis severity scores in citrus plants in three evaluations: in July 2009 (Cl, El and T1); in July 2010 (C2, E2 and T2) and in October 2010 (C3, E3 and T3).

This can be explained by the period in which the evaluation was carried out, for example, in July 2010, as this is a period of low rainfall, there is natural leaf fall due to the water deficit. Thus, visually, leprosis appears to have a higher severity, while in the evaluation carried out in October 2010 the plants were more leafy and had started to sprout in response to the rains, which may suggest that leprosis is at a lower level of severity.

It was observed that the transverse and longitudinal diameters differed between the collections, with the fruit from the last collection, in December 2010, showing the lowest values (Figure 17 A and B). The fruits from this collection (C3; E3; T3) refer to the second flowering in the 2009-2010 harvest, so these fruits competed with the fruits from the first flowering for photoassimilates for a relatively long time. In addition to this fact, due to some irregularities in the ripening of these second-flowering fruits, many of them had apparently not yet reached the point where they could be harvested.

Despite this, the juice yield was very similar for all the collections, varying between 50 and 58% (Figure 17 C).

The titratable acidity was higher in all treatments in the December 2010 harvest (Figure 18 A). This result is possibly explained by the fact that the fruit was not yet fully ripe and therefore contributed to the increase in acidity. The acidity of citrus juice increases at the beginning of fruit development, remaining practically constant until the initial stage of ripeness. The acidity decreases as the fruit ripens, due to dilution by the increase in fruit size (SOUZA, 2009). In general, however, the soluble solids content was lower in the July 2010 harvest. It can be seen that the treatment with sulphocalcium syrup (C2) had the lowest soluble solids content and differed significantly from the other treatments with spirodiclofen/cyhexatin (E2) and without acaricide (T2) (Figure 18 B).

It can be seen that the *ratio* differed between the three collections, with the lowest levels being found in the collection carried out in December 2010. On the other hand, the highest *ratio* values were found in the July 2010 collection, with apromimated values of 14 (Figure 19 A).

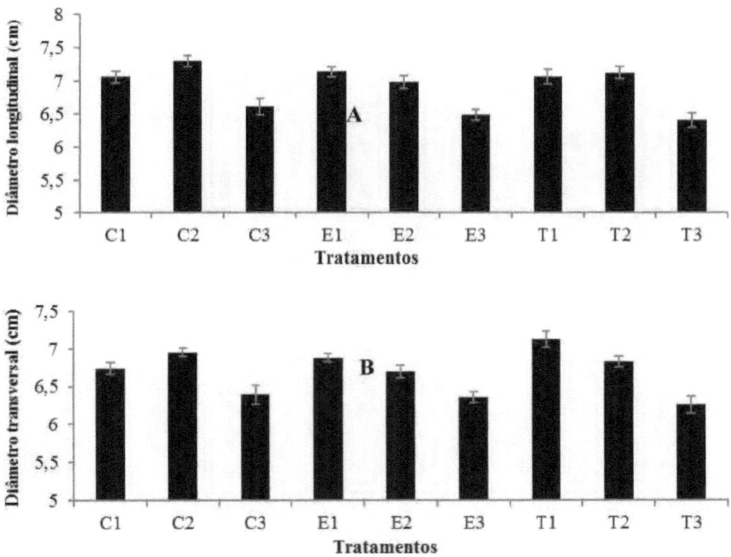

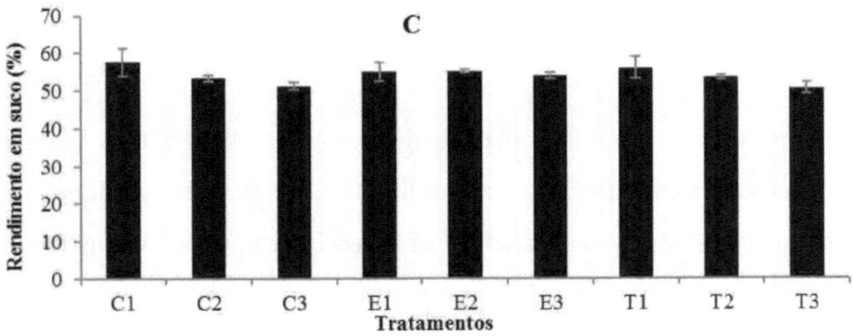

Figura 17. Average longitudinal (A) and transverse (B) diameters of citrus fruits and juice yield (C), obtained from three collections to evaluate physicochemical characteristics: in August 2009 (Cl, El and T1); in August 2010 (C2, E2 and T2) and in December 2010 (C3, E3 and T3).

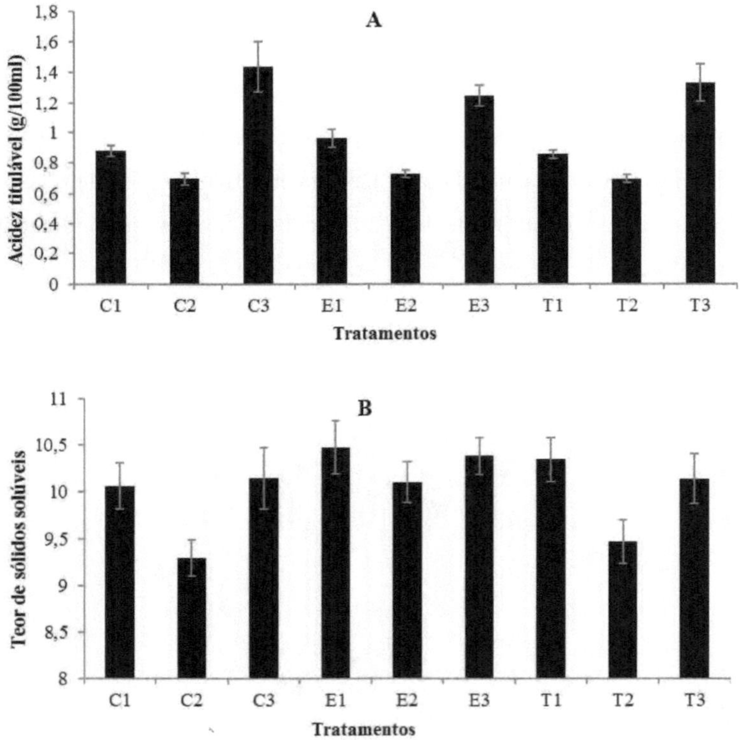

Figura 18. Average values of titratable acidity (A) and soluble solids content (Brix) (B) of citrus fruits obtained in three collections for the evaluation of physicochemical characteristics: in August 2009 (C1, E1 and T1); in August 2010 (C2, E2 and T2) and in December 2010 (C3, E3 and T3).

The technological index was slightly higher for fruit collected in July 2009. The industrial yield is given by the technological index, represented by the physical and chemical characteristics of the fruit, while the method used to determine the maturity and harvest time of orange fruit is the ratio between the percentages of total soluble solids and titratable acidity, known as the maturity index or *ratio*. As well as being an indicator of maturity, the technological index can also be used as an indicator of fruit quality (SOUZA, 2009; SOUZA & GOES, 2010).

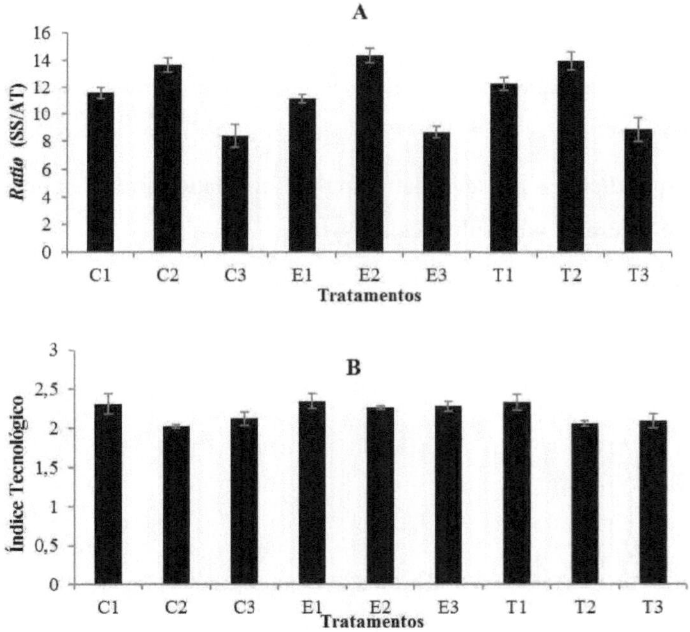

Figura 19. Average values of *ratio* (A) and technological index (B) of citrus fruits obtained from three collections for the evaluation of physicochemical characteristics: in August 2009 (C1, E1 and T1); in August 2010 (C2, E2 and T2) and in December 2010 (C3, E3 and T3).

5. CONCLUSIONS

- Fruits from plants with high leprosis severity that remain on the plant until harvest have identical physico-chemical characteristics to fruits from leprosis-free or low severity plants.

- There is no significant correlation between the severity of leprosis and the physico-chemical characteristics of the fruit: size, mass, juice yield, titratable acidity, soluble solids and *ratio*;

- Applications of the acaricides spirodiclofen, cyhexatin and sulphocalcic syrup did not significantly influence the parameters: size, mass, juice yield, titratable acidity, total soluble solids and *ratio*;

- Treatment with spirodiclofen and cyhexatin applied in rotation provides better control of leprosis than treatment with sulphocalcic syrup;

- The highest total production was seen in the treatment with spirodiclofen and cyhexatin;

- The physicochemical characteristics evaluated vary according to the fruit collection period;

6. SUMMARY

Despite the great economic and social importance of citrus leprosis in the Brazilian citrus agro-ecosystem, studies on the direct and indirect effects of this disease on fruit characteristics are still scarce. The aim of this work was to study the physico-chemical characteristics of fruit in citrus plants, with and without chemical control of the mite *Brevipalpus phoenicis*, in the management of citrus leprosis. The experiment was carried out at Fazenda Sao Pedro, located in the municipality of Reginópolis, SP. The experimental design was randomized blocks, with three treatments and 16 replications. The treatments used were: (1) spirodiclofen and cyhexatin applied in rotation; (2) sulfocalcic syrup; (3) no acaricide applied. Each experimental plot consisted of 3 plants arranged in a row. The acaricides were applied from the beginning of the experiment, and the differences in leprosis severity between the treatments became marked in the last harvests of the experiment. Spirodiclofen and cyhexatin were used at dosages of 20 and 50 mL p.c./100 L of water, respectively, while sulphocalcic syrup was used at a dosage of 4,000 mL p.c./100 L of water. Before harvesting, ten fruits were randomly collected from the central plant of each plot in August 2009, August and December 2010. After collecting the fruit, the samples were sent to the Chemical Analysis Laboratory at the Bebedouro Citriculture Experimental Station (EECB) to determine the fruit's physical and chemical characteristics. The fruit characteristics evaluated were size, mass, juice yield, titratable acidity, soluble solids, *ratio* and technological index. The results showed that there was no significant correlation between the severity of leprosis and these physical and chemical variables, nor did the application of acaricides significantly influence the variables evaluated. The treatment with spirodiclofen and cyhexatin applied in rotation provided better control of leprosis than the treatment with sulphocalcium syrup. The highest total production was seen in the treatment with spirodiclofen and cyhexatin. It was found that the physical and chemical characteristics evaluated varied according to the period during which the fruit was collected. Finally, it was found that fruit from plants with high leprosis severity that remain on the plant until harvest have identical physical and chemical characteristics to fruit from plants free of leprosis or with low severity.

7. SUMMARY

Despite the important economic and social leprosis citrus agroecosystem in Brazil, the studies on the direct and indirect effects of this disease on the characteristics of the fruits are scarce. The aim of this study was to study the physical and chemical characteristics of citrus fruit, with and without chemical control of the mite *Brevipalpus phoenicis*, in management of citrus leprosis. The experiment was installed at Sao Pedro's Farm, located in the city of Reginópolis, SP. The experimental design was randomized blocks with three treatments and 16 repetitions. The treatments used were: (1) spirodiclofen and cyhexatin applied in rotation, (2) lime sulfur, (3) without application of acaricide. Each plot consisted of 3 plants arranged in line. The applications of acaricides have been held since the beginning of the experiment, and differences in severity between treatments leprosis become marked in recent harvests of the experiment. The spirodiclofen cyhexatin and were employed at a dosage of 20 and 50 mL p.c./100 L, respectively, while the dosage of lime sulfur in 4000 mL p.c./100 L of water. Prior to harvest, ten randomly collected fruits from the central plant of each plot, and samples were collected in August 2009, August and December 2010. After the collection of fruit samples were sent to the Chemical Analysis Laboratory, the Experimental Station of Bebedouro - EECB for determining the physical and chemical characteristics of the fruit. The characteristics evaluated were fruit size, weight, juice yield, acidity, soluble solids, *ratio* and technology index. The results found no significant correlation between the severity of leprosis and these physicochemical parameters, and applications of acaricides in characteristics evaluated did not influence significantly. Treatment with

Spirodiclofen and cyhexatin applied in rotation and provided technological index superior to treatment with lime sulfur. The highest total production was observed in treatment and spirodiclofen and cyhexatin. It was found that the physical and chemical parameters measured show variations on the time period for collecting the fruits. Finally, it was found that the fruits of plants with high severity of lepers who remain on the plant until harvest have physical and chemical characteristics similar to the fruits of plants leprosis free or low severity.

8. REFERENCES

ABBOTT, C. E. The toxic gases of lime-sulfur. **Journal of Economic Entomology**, Lanham, v.38, n. 5, p.618-620, 1945.

AGRIANUAL 2009**: Yearbook of Brazilian Agriculture**. Sao Paulo: FNP Consultoria & Comércio, 2009.

AGUSTi, M. F.; ALMELA, V.; JUAN, M.; PRIMO-MILLO, E.; TRENOR, I.; ZARAGOZA, S. Effect of 3,5,6-trichloro-2-pyridyloxyacetic acid on fruit size and yield of 'Clausellina' mandarin (*Citrus unshiu* Marc.).

Journal of Horticultural Science, Ashford Kent, v.69, n.2, p.219-223, 1994.

ALBRIGO, G. Climatic influences on seasonal variation of Florida orange pounds solids. **Proceedings International Society Horticulturae Science**, Geneva, v.2, n. 6, p.15-18, 1990.

ALBUQUERQUE, F. A. de; OLIVEIRA C. A. L. de; BARRETO, M. Behavior of the mite *Brevipalpus phoenicis* (Geijskes, 1939) (Acari:Tenuipalpidae) on citrus fruits. In: OLIVEIRA, C. A. L. de; DONADIO, L. C. (Ed.). **Citrus leprosis**. Jaboticabal: FUNEP, 1995. p.77-90.

ALBUQUERQUE, F. A.; OLIVEIRA, C. A. L. de; BARRETO, M. Estudos da relação entre as incidências de verrugose da laranja-doce e leprose dos citros em frutos de laranja-pera. **Cientifica**, Sao Paulo, v.25, n.2, p.393-402. 1997.

ALVES, E. B.; CASARIN, N. F. B.; OMOTO, C. Dispersal mechanisms of *Brevipalpus phoenicis* Geijskes (Acari:Tenuipalpidae) in citrus groves **Neotropical Entomology**, Londrina, v.34, n.1, p.89-96, 2005.

ANDRADE, D. J.; OLIVEIRA, C. A. L. de; FALCONI, R. S.; PATTARO, F. C.; FERNANDES, E. J. Effect of water deficit and the presence of citrus leprosis virus on leaf temperature. **Laranja**, v.29, p.1 - 15, 2008.

ANDRADE, D. J.; OLIVEIRA, C. A. L.; FALCONI, R. S.; PATTARO, F. C.; FERNANDES, E.J. Effect of water deficit and presence of citrus leprosis virus on leaf temperature and its relationship with *Brevipalpus phoenicis* population and disease severity in citrus plants. **Laranja**, Cordeirópolis, v.29, n.1, p.1-15, 2008.

ANDRADE, D. J.; OLIVEIRA, C. A. L. de; PATTARO, F. C. Evaluation of the efficiency of acaricides, by direct and residual action, on *Tetranychus mexicanus* (McGregor, 1950) (Acari: Tetranychidae) in citrus. **Revista de Agricultura**, Piracicaba, v.83, n.1, p.28-34, 2008.

ANDRADE, D. J.; OLIVEIRA, C. A. L. de; ROMANI, G. N.; PATTARO, F. C. Efeito da calda sulfocâlcica sobre o acaro *Tetranychus mexicanus* (McGregor, 1950) em citros. **Revista de Agricultura**, Piracicaba, v.82, p.161-169, 2007.

ANDRADE, D. J. et al. Acaricides used in conventional and organic citriculture: management of leprosis and phytoseiid mite populations. **Revista Brasileira de Fruticultura**, Jaboticabal, v.32, p.1028-1037, 2010.

BABCZINSKI, P. Environmental behavior of spirodiclofen (BAJ 2740; Envidor). **Pflanzenschutz-Nachrichten Bayer**, Leverkusen, v.55, n.2-3, p.197-206, 2002.

BARTHOLOMEW, E. T.; SINCLAIR, W. B.; Soluble constituents and buffer properties of orange juice.**Plant Physiology**, Rockville, v.18, n.2, p.185- 206. 1943.

BASSANEZI, R. B. Aspectos da leprose dos citros. **Fitopatologia Brasileira,** Brasilia,v.26, p.246-247, 2001.

BASSANEZI, R. B. Citrus leprosis: focus on the control of the mite vector. **Visao Agricola.** Piracicaba, n.2, p.24-29, 2004.

BASSANEZI, R. B.; SPÓSITO, M. B.; YAMAMOTO, P. T. Farewell to leprosis. **Revista Cultivar**, Pelotas, v.10, n.1, p.37, 2002.

BASTIANEL, M.; FREITAS-ASTÙA, J.; RODRIGUES, V.; ANTONIOLI-LUIZON,R.; ARRIVABEM, F.; MACHADO, M. A. Response of 'murcott' tangor to citrus leprosis virus inoculation in field and greenhouse. **Laranja**, Cordeirópolis, v.25, n.2, p.337-348, 2004.

BASTIANEL, M.; FREITAS-ASTÙA, J.; KITAJIMA, E. W.; MACHADO, M. A. The citrus leprosis pathosystem.**Summa Phytopathologica**, Botucatu, v.32, n.3, p.211-220, 2006.

BASTIANEL, BASTIANEL, M.; NOVELLI, V. M.; KITAJIMA, E. W. ; KUBO, K. S.; BASSANEZI, R. B.; MACHADO, M. A.; FREITAS- ASTÙA, J. Citrus Leprosis: Centennial of an Unusual Mite Virus Pathosystem. **Plant Disease,** v.94, p.284-292, 2010.

BERTOLDO, A. A. **Development of a methodology for determining the components and specifying polysulphides in samples of sulphocalcic grout**. 2003, 87f. Dissertation (master's degree in

Quimica) Instituto de Quimica, Universidade de Sao Paulo, Sao Carlos, 2003.

BITANCOURT, A. A. A leprose e a próxima colheita. **O Biológico**, Sao Paulo, v.3, n.1, p. 37-40, 1937.

BITANCOURT, A. A. Studies on citrus leprosis IV**. Arquivos do Instituto Biológico**, Sao Paulo, v.22, n.1, p.219-231, 1955.

BUSLIG, B. S. The orange. In: ESKIN, N. A. M. (Ed.). **Quality and preservation of fruits.** Boca Raton: CRC Press, 1991. chap. 1, p.1-15.

BUSOLI, A. C. O manejo integrado de pragas - citros e a busca de qualidade total na citricultura. **Laranja**, Cordeirópolis, v.16, n. 1, p.155186, 1995.

CALEGARIO, R. F. ; KITAJIMA, E. W. ; SALAROLI, R. B. ; REZENDE, J. A. M.; LOCALI-FABRIS, E. C. Immunofluorescence studies on the distribution of citrus leprosis virus (CiLV-C) antigen in citrus leaf lesions. In: XVIII Conference Intl.Org.Citrus Virologists, 2010, Campinas, SP. Citrus Research and Technology. Cordeiropolis, SP: Centro APTA Citros Sylvio Moreira, Inst. Agronômico, 2010. v.31. p.5555.

CAMARGO, M. B. P.; PEDRO JÛNIOR, M. J.; ORTOLANI, A. A. ROSA, J. M. Development and testing of agrometeorological models for estimating the productivity of orange groves in the state of Sao Paulo. In: CONGRESSO BRASILEIRO DE AGROMETEOROLOGIA, 9., Campina

Grande, 1995. **Proceedings...** Campina Grande: Brazilian Society of Agrometeorology, 1995. p.412-414.

CASARIN, N. F. B. **Sulfocaecic syrup in citrus orchards: evolution of resistance in *Brevipalpus phoenicis* (Acari: Tenuipalpidae) and impact on *Iphiseiodes zuluagai* (Acari: Phytoseiidae)**. 2010. 94f. Thesis (PhD in Entomology) - Escola Superior de Agricultura "Luiz de Queiroz", Universidade de Sao Paulo, Piracicaba, 2010.

CATI. **Citrus:** recommendations for the control of the main pests and diseases in orchards in the State of São Paulo**.** Campinas, 1997, p.58 (Technical Bulletin, 165).

CHIAVEGATO, L. G.; SALIBE, A. A. Damage caused by the mite *Brevipalpus phoenicis* (Geijkes, 1939) (Acari: Tenuipalpidae) on fruits of different citrus varieties. In: CONGRESSO BRASILEIRO DE FRUTICULTURA, 6, Recife, 1981. *Proceedings*... Recife: SBF, 1981. V.2, P. 709-718.

CHIAVEGATO, L. G. Evaluation of the potential of *Brevipalpus phoenicis* (Geijskes, 1939) (Acari: Tenuipalpidae) in the transmission of leprosis in citrus plants. In: CONGRESSO BRASILEIRO DE ENTOMOLOGIA, 15, 1995. Caxambu, **Anais**..., p.14.

CHIAVEGATO, L. G. Biology of the mite *Brevipalpus phoenicis* in citrus. **Pesquisa Agropecuâria Brasileira**, Brasila, v.21, n.8, p.813-816, 1986.

CHIAVEGATO, L. G.; MISCHAN, µ. M.; SILVA, µ. A. Damage and transmissibility of leprosis symptoms by the mite *Brevipalpus phoenicis* (Geijskes, 1939) Saied, 1946 (Acari, Tenuipalpidae) in citrus. **Cientifica,** Sao Paulo, v.10, n.2, p.265-271, 1982.

CHILDERS, C. C.; FRENCH, J. V.; RODRIGUES, J. C. V. *Brevipalpus californicus, B. obovatus, B. phoenicis* and *B. lewisi* (Acari:Tenuipalpidae): a review of theirbiology, feeding injury and economic importance. **Experimental and AppliedAcarology**, Amsterdam, v.30, n.1-3, p.5-28, 2003a.

CHILDERS, C. C.; RODRIGUES, J. C. V.; WELBOURN, W. C. Host plants ofBrevipalpus *californicus, B. obovatus,* and *B. phoenicis* (Acari:Tenuipalpidae) and their potential involvement in the spread of one or more viral diseases vectored by these mites. **Experimental and Applied Acarology**, Amsterdam, v.30, n.1-3, p.29-105, 2003b.

COLARICIO, A.; LOVISOLO, O.; CHAGAS, C. M.; GALLETI, S. R.; ROSSETI, V.; KITAJIMA, E. W. Mechanical transmission and ultrastructural aspects of citrus leprosis desease. **Fitopatologia brasileira**, Brasilia, v.20, p.208-213, 1995.

CORBETT, J. R.; WRIGHT, K.; BAILLIE, A. C. **The biochemical mode of action of pesticides**. London: Academic Press, 1984, 382p.

CRUSE R. R.; WIEGAND C. L.; SWANSON W. A.The effects of rainfall and irrigation management on citrus juice quality in Texas.**Journal of AmericanSociety of Horticultural Science**, Alexandria.v.107, n.5, p.767770, 1982.

CUNHA SOBRINHO, A. P.; SOARES FILHO, W. S.; PASSOS, O. S.; CALDAS, R. C. Influence of rootstocks on fruit quality of Iaranjeira 'Baianinha' under tropical conditions. **Revista Brasileira de Fruticultura**, Cruz das Almas, v.14, n.3, p.99-104, 1992.

DI GIORGI, F.; YDE, B. Y.; TRIBONI, H. R.; MARCHI, R. J.; WAGNER, R. L. Quality of oranges for industrialization. **Laranja,** Cordeirópolis, v.14, n.1, p.97-118, 1993.

DI GIORGI, F; IDE, B. Y.; DIB, K.; MARCHI, R. J.; TRIBONI, H. R.; WAGNER, R. L.; ANDRADE, G. Influence of climate on orange production. **Laranja,** Cordeirópolis, v.12, n.1, p.163-192, 1991.

DOMINGUEZ, F. S.; BANDEL, A.; CHILDERS, C.; KITAJIMA, E. W. First reportof citrus leprosis on Panama. **Plant Disease,** Saint Paul, v.85, n.2, p.228, 2001.

ERICKSON, L. E. The general physiology of citrus. In: REUTHER, W.; BATCHELOR, L. D.; WEBBER, H. J. (Ed). **The citrus industry.** Riverside: University of California, 1968. v.2, p.86-122.

FAWCETT, H. S. Scaly bark or nail-head rust of citrus. Bull.**Florida Agricultural Experimental Station Bulletin**, Gainesville, v.106, p.41, 1911.

FEICHTENBERGER, E. Integrated management of the main citrus diseases in Brazil. In: SEMINARIO INTERNACIONAL DE CITROS, 6. 2000, Bebedouro. **Proceedings...**, p.177-216, 2000.

FEICHTENBERGER, E.; MÜLLER, G. W.; GUIRADO, N. *Citrus* diseases (*Citrus spp*). In: KIMATI, H.; AMORIN, L.; FILHO, A. B.; CAMARGO, L. E. A.; REZENDE, J. A. M. **Manual de fitopatologia**: doenças das plantas cultivadas. Sao Paulo: Agronòmica Ceres, v.2, 1997, p.261-296.

FERNANDES, A. P.; OLIVEIRA, C. A. L. de; FERREIRA, M. C. Efficiency of acaricides on populations of *Brevipalpus phoenicis* (Geijskes) (Acari Tenuipalpidae) of citrus and coffee trees. **Revista de Agricultura**, Piracicaba, v.83, n.1, p.28-34, 2008.

FISCHER, R.; BENET-BUCHHOLZ, J. Chemistry and stereochemistry of spirodiclofen (BAJ 2740). **Pflanzenschutz-Nachrichten Bayer**, Leverkusen, v.55, n.2-3, p.1137-147. 2002.

FRANCO C. R.; CASARIN, N. F. B.; DOMINGUES, F. A.; OMOTO, C. Resistance of *Brevipalpus phoenicis* (Geijskes) (Acari: Tenuipalpidae) to acaricides that inhibit cellular respiration in citrus: Cross-resistance and adaptive cost. **Neotropical Entomology**, Londrina, v.36, n.4, p.565576, 2007.

FREITAS-ASTÙA, J.; BASTIANEL, M.; LOCALI-FRABRIS, E. C.; NOVELLI, V. M.; SILVA-PINHATI, A. C.; PALMIERI, A. C.; TARGON, M. L. P. N.; MACHADO, M. A. Differentially expressed stress-related genes in the compatible citrus-Citrus leprosis virus interaction. **Genetic Molecular Biology**, Sao Paulo, v.30, n.3, p.980-990, 2007.

FREITAS-ASTÙA, J.; BASTIANEL, M.; LOCALI-FRABRIS, E. C.; NOVELLI, V. M.; SILVA-PINHATI, A. C.; PALMIERI, A. C.; TARGON, M. L.; MACHADO, M. A. Differentially expressed stress- related genes in the compatible Citrus-Citrus Ieprosis virus interaction. **Genetic Molecular Biology**, Sao Paulo, v.30, n.3, p. 980-990, 2007.

FREITAS-ASTÙA, J.; FADEL, A. L.,; BASTIANEL, M.; NOVELLI, V. M.;

ANTONIOLI-LUIZON, R.; MACHADO, M. A. Differential response of citrus species and hybrids to leprosis. **Pesquisa Agropecuària Brasileira**, Brasilia, v.43, p.809-814. v.43, n.7, p.809-814, 2008.

GUARDIOLA, J. L. Flowering and productivity of citrus. **Laranja & Cia**, Matao, v.33, jul./set., p.4-6, 1992.

GOLDSCHMIDT, E.E.; GOLOMB, A. The carbohydrate balance of alternate bearing citrus trees and the significance for flowering and fruiting. **Journal of American Society for Horticultural Science**, Ashford, v. 107, p. 206-208, 1982.

HARAMOTO, F. H. Biology and control of *Brevipalpus phoenicis* (Geijskes, 1939) (Acari: Tenuipalpidae). **Hawaii Agricultural Experimental Station Technical Bulletin**, Honolulu, n.68, p.1-60, 1969.

HELLE, W.; BOLLAND, H. R. & HEITMANS, W. R. B. Chromosomes and types ofparthenogenesis in false spider mites (Acari: Tenuipalpidae). **Genetica**, Netherlands, v.54, n.1, p.45-50, 1980.

HOFFMANN, H. Organic food and artisanal products. **Visao Agricola**. Piracicaba, n.2 , p.24-29, 2004. Available at: < http://www.saude.sc.gov.br>. Accessed on: October 14, 2004.

HULL, R. Matthews' Plant Virology. 4th edition. **Academic Press**, Orlando, 1001p. 2002.

IRAC. Insecticide Resistance Action Committee. **Advances in the implementation of pesticide pest resistance management programs in Brazil**. Available at: <http://www.irac-r.org.br/arquivos/avancosimplprog.doc>. Accessed on: 02 Nov. 2010.

KITAJIMA, E. W.; GROOT, T. V. M.; NOVELLI, V. M.; FREITAS- ASTÙA, J.;ALBERTI, G.; MORAES, G. J. de. In situ observation of the *Cardinium* symbiontsof *Brevipalpus* (Acari: Tenuipalpidae) by electron microscopy. **Experimental and Applied Acarology**, Amsterdam, v. 42, p. 263-271, 2007.

KITAJIMA, E. W.; LOVISOLO, O.; COLARICCIO, A.; CHAGAS, C. M.;

ROSSETTI, V. Virus causing citrus leprosis. In: OLIVEIRA, C. A. L.; DONADIO, L. C. (ed.). **Citrus leprosis**. Jaboticabal, FUNEP: 1995. p. 19-32.

KITAJIMA, E. W.; MÜLLER, G. W.; COSTA, A. S.; YUKI, V. A. Short, rod like particles associated with citrus leprosis. **Virology**, Orlando, v.50, n.1, p.254-258, 1972.

KITAJIMA, E. W.; CHAGAS, C. M.; RODRIGUES, J. C. V. Brevipalpus- transmited plant virus and virus-like diseases: cytopathology and some recent cases. **Experimental and Applied Acarology**, Amsterdam, v.30, n.1-3, p.135-160, 2003.

KITAJIMA, E. W.; FERREIRA, P. T. O.; FREITAS-ASTÙA, J.; MACHADO, M. A. Occurrence of citrus leprosis, nuclear type (CiLV-N), in the São Paulo municipalities of Monte Alegre do Sul and Amparo. **Summa Phytopathologica**, Botucatu, v.30, n.1, p.68, 2004.

KITAJIMA, E. W.; ROSILLO, M. A.; PORTILLO, M. M.; MULLER, G. W.; COSTA, A. S. Electron microscopy of leaf tissues of orange trees infected by Argentine explosive leprosy. **Fitopatologia**, Lima, n.9, p.55-56, 1974.

KNORR, L. C. Studies on the etiology of leprosis in citrus. In: CONFERENCE OFTHE INTERNATIONAL ORGANIZATION OF CITRUS VIROLOGISTS, 4., 1968, Gainnesville. Proceedings... Gainnesville: IOCV,1968, p. 332-341.

KONNO, R. H.; FRANCO, C. R.; OMOTO, C. Susceptibility of populations of *Brevipalpus phoenicis* (Geijskes, 1939) (Acari: Tenuipalpidae) to organotin acaricides in citrus. **Scientia Agricola**, Piracicaba, v.58, n.4, p.703-709, 2001.

LARANJEIRA, F. F.; AMORIM, L.; BERGAMIN FILHO, A.; AGUILAR-VILDOSO, C. I.; DELLA COLLETA FILHO, H. Fungi, prokaryotes and abiotic diseases. In: MATTOS JÙNIOR, D.; DE NEGRI, J. D.; PIO, R. M.; POMPEU JÙNIOR, J. (Ed.). **Citrus**. Campinas: IAC, 2005, p.147-195.

LOCALI-FABRIS, E. C.; FREITAS-ASTUA, J.; SOUZA. A. A. TOKITA, M. A., ASTUA-MONGE, G.; ANTONIOLI-LUIZON, R.; RODRIGUES, V.; TARGON, M. L. P. N.; MACHADO, M. A. Complete nucleotide sequence, genomic organization and phylogenetic analysis of Citrus Leprosis virus cytoplasmatic type. **Journal**

General Virology, London, v.87, n. 1, p. 2721-2729, 2006.

MAIA, O. M. A.; OLIVEIRA, C. A. L. Colonization capacity of *Brevipalpus phoenicis* (Geijskes) (Acari: Tenuipalpidae) on hedges, windbreaks and invasive plants. **Neotropical Entomology**, v.33, n.5, p.625-629, 2004.

MAIA, O. M. A.; OLIVEIRA, C. A. L. Transmission of leprosis virus from hedges, windbreaks and weeds to orange trees by *Brevipalpus phoenicis* (Geijskes). **Bragantia**, Campinas, v.64, n.3, p.417422, 2005.

MARQUES, J. P. R. ; KITAJIMA, E. W. ; FREITAS-ASTUA, J. ; APPEZATO-DA-GLORIA, B. Comparative morpho-anatomical studies of the lesions caused by citrus leprosis virus on sweet orange. **Anais da Academia Brasileira de Ciências (Impresso)**, v.82, p.501-511, 2010.

MARTINELLI, N. M.; OLIVEIRA, C. A. L.de; PERECIN, D. Basic knowledge for studies involving surveys of the population of the mite *Brevipalus phoenicis* (Geijskes,1939) in citrus cultivation. **Cientifica**, Sao Paulo, v.4, p.242-253, 1976.

MEDINA, C. L.; MACHADO, E. C.; RENA, A. B.; SIQUEIRA, D. L. de. Citrus physiology. In: MATTOS JÙNIOR, D.; DE NEGRI, J. D.; PIO, R. M.; POMPEU JÙNIOR, J. (Ed.). **Citrus**. Campinas: IAC, 2005, p.147195.

MESA, N. C.; OCHOA, R.; WELBOURN, W. C.; EVANS, G. A.; MORAES, G. J. 2009. A catalog of the tenuipalpid (Acari) of the world with a key to genera.ed. Auckland: Zootaxa, **Magnolia Press**, 2009. v.1, 185p.

MURAYAMA, D.; AGRAWAL, H. O.; INOUE, T.; KIMURA, I.; SHIKATA, E.; TOMARU, K.; TSUCHIZAKI, T.; TRIHARSO (Ed.). Plant Viruses in Asia.Bulasksumur: **Gadjah Mada University Press**, 2000, 1263p.

NAKANO, O.; SANCHEZ, G. A.; ISHIDA, A. K. Reduction in the infestation of the leprosis mite *Brevipalpus phoenicis* (Geijskes, 1939) in citrus through the control of verrucas. **Orange**. Cordeirópolis, v.8, n. 1, p.1933, 1987.

NUNES, M. A. Verification of the transmissibility of citrus leprosis virus by *Brevipalpus phoenicis* (Geijskes, 1939) in hedges, windbreaks and weeds using transmission electron microscopy. 2004. 46 f. **Dissertation** (Master's degree in Agricultural Entomology), Universidade Estadual Paulista, Jaboticabal, 2004.

OLIVEIRA, C. A. L. de. Living fences, windbreaks and weeds in leprosis management. **Visâo Agricola**. Piracicaba, n.2, p.30-31, 2004.

OLIVEIRA, C. A. L. de; PATTARO, F. C. Effect of the acaricide spirodiclofen on the fecundity, fertility and survival of females *of Brevipalpus phoenicis* (Acari: Tenuipalpidae) in the laboratory. In: CONGRESSO BRASILEIRO DE ENTOMOLOGIA, 20., 2004a, Gramado...**Abstracts**... p. 169.

OLIVEIRA, C. A. L. de; PATTARO, F. C. Leprosis: Management strategies that can eradicate it from your orchard. Production Technologies. HFF & Citrus, Jaguariûna, v.3, n. 3, p.34-36, 2004b.

OLIVEIRA, C. A. L. de; PATTARO, F. C. Management of citrus leprosis. In: COURSE ON CITRUS DISEASES AND THEIR MANAGEMENT, 1., **CD-ROM**, 2005, Cordeirópolis.

OLIVEIRA, C. A. L. de; SANTOS, J. R. J. E.; SALA, I. Citrus leprosis mite: results of 104 field trials aimed at its control 1985-1990.Jaboticabal: FUNEP, 1991,56p.

OLIVEIRA, C. A. L. Population fluctuation and control measures of the *miteBrevipalpus phoenicisin* citrus. **Laranja**, Cordeirópolis, v.6, n. 1, p.1-32, 1986.

OLIVEIRA, M. L.; PATTARO, F. C.; OLIVEIRA, C. A. L. de. Influence of the residue of sulfocaecic syrup on the efficiency of Vertimec 18 CE in the control of *Phyllocoptruta oleivora*, in citrus. In: CONGRESSO BRASILEIRO DE ENTOMOLOGIA, 19., 2002, Manaus. **Abstracts**...p. 252.

OLIVEIRA, R. M.; VENZON, M.; FREITAS, R. C. P.; GIRALDO, A. S.; OLIVEIRA, H. G.; PALINI, A. Efficiency of sulphocalcic syrup in the field in the control of the coffee red mite *Oligonychus ilicis*. In: SIMPÓSIO BRASILEIRO DE ACAROLOGIA, 2., 2008, Poços de Caldas, **Abstracts**... p. 119, 2008.

OMOTO, C. Resistance of *Brevipalpus phoenicis* (Acari: Tenuipalpidae) to chemical products in citriculture. In: OLIVEIRA, C. A. L. de;

DONADIO, L. C. **Citrus leprosis**. Jaboticabal-SP: FUNEP, 1995, p.179-188.

PASCHOLATI, S. F. Phytopathogens: phytotoxins and hormones. In: BERGAMIN FILHO, A.; KIMATI, H. & AMORIN, L. (Ed). Manual de fitopatologia, 3.ed., Sao Paulo: **Agronòmica Ceres**, 1995, v.1, p.365-392.

PASCON, R. C.; KITAJIMA, J. P.; BRETON, M. C.; ASSUMPÇÂO, L.; GREGGIO, C.; ZANCA, A. S.; OKURA, V. K.; ALEGRIA, M. C.; CAMARGO, M. E.; SILVA, G. G. C; CARDOZO, J. C.;VALLIM, M. A.; FRANCO, S. F.; SILVA, V. H.; JORDÂO JÙNIOR, H.; OLIVEIRA, F.; GIACHETTO, P. F.; FERRARI, F.; AGUILAR-VILDOSO, C. I.; FRANCHISCINI, F. J. B.; SILVA, J. M. F.; ARRUDA, P.; FERRO, J. A.; REINACH, F.; SILVA, A. C. R.. The complete nucleotide sequence and genomic organization of Citrus Leprosis associated Virus, Cytoplasmatic type (CiLV-C). **Virus Genes**, New York, v.32, n. 3, p.289-298, 2006.

PATTARO, F. C. Sulfocálcica Calda no Agrossistema Citricola. 2003. 73 f. **Dissertation** (Master's Degree in Agricultural Entomology) - Faculty of Agricultural and Veterinary Sciences, São Paulo State University, Jaboticabal, 2003.

PATTARO, F. C. Technical and economic aspects of pruning and chemical control in the management of citrus leprosis. 2006. 140p. **Thesis** (Doctorate in Agricultural Entomology) - Faculty of Agricultural and Veterinary Sciences, Paulista State University, Jaboticabal, 2003.

PIJNACKER, L. P.; FERWERDA, M. A.; BOLLAND, H. R.; HELLE, W. Haploidfemale parthenogenesis in the false spider mite *Brevipalpus obovatus* (Acari: Tenuipalpidae). **Genetica**, Bethesda, v.51, n.3, p.211-214, 1980.

PIJNACKER, L. P.; FERWERDA, M. A.; HELLE, W. Cytological on the female and male reproductive system of the parthenogenetic privet mite *Brevipalpus obovatus* Donnadieu (Phytoptipalpidae, Acari). **Acarologia**, Prasad, v.22, n.2, p.157-163, 1981.

ORGANIC PLANET. Organic citriculture - I. Available at <http://www.planetaorganico.com.br>Accessed on January 14, 2003.

POLETTI, M.; OMOTO,C. Cross-resistance relationships between spirodiclofen and some acaricides recommended for the management of *Brevipalpus phoenicis* (Acari: Tenuipalpidae) in citrus. In: CONGRESSO BRASILEIRO DE ENTOMOLOGIA, 20., 2004, Gramado-RS. **Abstracts**... Gramado: SEB, 2004, p.515.

POZZAN, M.; TRIBONI, H. R. Harvesting and fruit quality. In: MATTOS JÙNIOR, D.; NEGRI. J. D.; PIO, R. M.; POMPEU JÙNIOR, J. (Ed.). **Citrus**. Campinas: Instituto Agronômico; Fundag, 2005, p. 800-822.

PRATES, H. S. Bordeaux mixture, sulphocalcium and viçosa - alternative products in citriculture. 1999. Available at <http://www.cati.sp.gov.br>Accessed June 2008.

REDD, J. B.; HENDRIX JUNIOR. C. M.; HENDRIX, D. L. Quality control manual for citrus processing plants book. I - Regulation, citrus methodology, microbiology, conversion charts, tables, others. Florida, Intercity, 1986.250p.

REIS, P. R.; NETO, M. P.; FRANCO, R. A.; TEODORO, A. V. Control of *Brevipalpus phoenicis* (Geijskes, 1939) and *Oligonychus ilicis* (Mcgregor, 1917) (Acari: Tenuipalpidae, Tetranychidae) in coffee trees and the impact on beneficial mites. II - spirodiclofen and azocyclotin. **Ciência e Agrotecnologia**, Lavras, v.29, n.3, p.528-537, 2005.

REIS, P. R.; SOUZA, J. C.; SOUSA, E. O.; TEODORO, A. V. Control of the mite *Brevipalpus phoenicis,* vector of coffee leaf spot. In: CONGRESSO BRASILEIRO DE ENTOMOLOGIA, 17.,1998, Rio de Janeiro, **Abstracts**... v.2, p. 1502.

REUTHER, W. Climate and citrus behavior. In: REUTHER, W. (Ed.). The citrus industry.Berkeley: University of California, 1973, v.3, p.280-337.

RODRIGUES, J. C. V. Citrus leprosis: cytopathology, transmissibility and relationship with the vector *Brevipalpus phoenicis* Geijskes (Acari: Tenuipalpidae).1995. 79f **Dissertation** (Master of Science) - Center for Nuclear Energy in Agriculture, University of Sao Paulo, Piracicaba, 1995.

RODRIGUES, J. C. V. Citrus leprosis: vector x pathogen relationship. In: OLIVEIRA, C. A. L. de; DONADIO, L. C. **Citrus leprosis**. Jaboticabal-SP: FUNEP, 1995. p.57-68.

RODRIGUES, J. C. V.; KITAJIMA, E. W.; CHILDERS, C. C.; CHAGAS, C. M. Citrus leprosis virus vectored by *Brevipalpus phoenicis* (Acari: Tenuipalpidae) on citrus in Brazil. **Experimental Applied Acarology**, Amsterdam, v.30, p.161-179, 2003

RODRIGUES, J. C. V. Program to reduce the inoculum of citrus leprosis. Laranja, Cordeirópolis, v.23, n.2, p.307-332, 2002.

RODRIGUES, J. C. V. Pathogen-vector-plant relationships in the citrus leprosis system, 2000. 169 f., **Thesis** (Doctorate) - Center for Nuclear Energy in Agriculture, University of Sao Paulo, Piracicaba, 2000.

RODRIGUES, J. C. V.; GALLO-MEAGHER, M.; OCHOA, R.; CHILDERS, C. C.; ADAMS, B. J. Mitochondrial DNA and RAPD polymorphisms in the haploid miteBrevipalpus *phoenicis*(Acari: Tenuipalpidae).**Experimental and Applied Acarology**, Amsterdan,v.34, p.275-90, *2004.*

RODRIGUES, J. C. V.; CHILDERS, C. C.; KITAJIMA, E. W.; MACHADO, M. A.; NOGUEIRA, N. L. A strategy for the control of citrus leprosis. **Laranja**, *Cordeirópolis, v.22, n.2, p.411-423, 2001.*

RODRIGUES, J. C. V.; KITAJIMA, E. W.; CHILDERS, C. C.; CHAGAS, C. M. Citrus leprosis virus vectored by Brevipalpus phoenicis (Acari: Tenuipalpidae) on citrus in Brazil. **Experimental and Applied Acarology**, *Amsterdam, v.30, n.1, p.161-179, 2003.*

RODRIGUES, J. C. V.; MULLER, G. W.; NOGUEIRA, N. L.; MACHADO, M. A. Yield damage associated to citrus leprosis on sweetorange varieties. **In: Proceedings of the International Society of Citriculture**. *Orlando, v.1, p.1055-1059, 2000.*

ROESSING, C.; SALIBRE, A. A. Incidence of leprosis in citrus *varieties*. **Ciência e Cultura**, Campinas, v.19, p.303, 1967.

ROSSETTI, V. V. Illustrated manual of citrus diseases: diseases caused by viruses and the like. Piracicaba: Fealq/Fundecitrus, 2001, p.207.

SALVA, R. A.; MASSARI, C. A. Situation of the leprosis mite in the State of São

Paulo - Fundecitrus survey. In: OLIVEIRA, C. A. L. de; DONADIO, L.C. Leprose dos citros. Jaboticabal: FUNEP, 1995, p. 13-17.

SILVA, J. A. A. Behavior of the 'Pêra' orange tree on two rootstocks and three levels of irrigation. 1999. 117 f. **Thesis** (PhD in Plant Production). Universidade Estadual Paulista, Jaboticabal, 1999.

SINCLAIR, W. B. Principal juice constituents In: SINCLAIR, W. B. The orange: its biochemistry and physiology. Riverside: University of California, 1960. chap. 5, p. 131-160.

SOUZA, P. F. C. Evaluation of sweet orange trees in terms of fruit quality, ripening periods and resistance to *Guignardia citricarpa.* 2009. 102p. **Thesis** (Doctorate in Agronomy) - Faculty of Agricultural and Veterinary Sciences, Universidade Estadual Paulista, Jaboticabal, 2009.

SOUZA, P. F. C.; GOES, A. de. Reaction of sweet orange trees regarding resistance to Guignardia citricarpa. **Revista Brasileira de Fruticultura**, Jaboticabal-SP, v.32, n.3, p.718-725, 2010.

SPEGAZZINI, C. Sobre algunas enfermedades y hongos que afectan las plantas de "agrios" en el Paraguay. **Annales de la Sociedad Cientifica**, Argentina, Buenos Aires, v.90, n.1, p.155-188, 1920.

TING, S. V. Citrus fruits In: CHA, H. T. J. Handbook of tropical foods. New York: Marcel Dekker, 1983. chap. 5, p.201-253.

TING, S. V.; ROUSEFF, R. L. Citrus product technology In: TING, S. V.; ROUSSEF, R. L. Citrus fruits and their products: analysis technology. New York: Marcel Dekker, 1986. chap. 2, p.7-16.

TRINDADE, M. L. B.; CHIAVEGATO, L. G. Biological characterization of the mites *Brevipalpus obovatus* D., *B. californicus* B. and *B. phoenicis* G. (Acari: Tenuipalpidae). **Anais da Sociedade Entomológica do Brasil**, Itabuna, v.23, n.2, p.189-195, 1994.

TUBELIS, A.; SALIBE, A. A. Relationship between 'Hamlin' orange productivity on

'Caipira' orange rootstock and monthly rainfall in the Botucatu highlands. **Pesquisa Agropecuâria Brasileira**, Brasilia, v.23, n.3, p.239-246, 1998.

VERGANI, A. R. Transmisión y naturaleza de la lepra explosiva del naranjo. (Publ.).Ministerio Agriculturade la Nación Buenos Aires/Instituto Sanidad Vegetal, Buenos Aires, v.3, n. 3, p.11, 1945.

VICENTE, C. A. Foliar application of chelated calcium to citrus fruit. 1990. 133 f. **Thesis** (PhD in Plant Production), Universidade Estadual Paulista, Jaboticabal, 1990.

WACHENDORFF, U.; NAUEN, R.; SCHNORBACH, H. J.; RAUCH, R.; ELBERT, A. The biological profile of spirodiclofen (Envidor) - A new selective tetronic acid acaricide. **Pflanzenschutz-Nachrichten**, Bayer, Leverkusen, v.55, n.2-3, p.149-163, 2002.

WEEKS, A.; MAREK, F.; BREEUWAR, J. A. J. A mite species that consists entirely of haploid females.**Science**, Washington, v.292, p. 24792482, 2001.

WEEKS, A. R.; VELTEN, R.; STOUTHAMER, R. Incidence of a new sex *ratio* distorting edosymbiotic bacterium among arthropods.**Proceedingsof the Royal Society of London, B,** London,v.270, p.1857-1865, 2003.

WELBORN, W. C.; OCHOA, R.; KANE, E. C.; ERBE, E. F. Morphologicalobservations on *Brevipalpus phoenicis* (Acari: Tenuipalpidae) including comparisons with *B. californicus* and *B. obovatus*. **Experimental and Applied Acarology**, Amsterdam, v.30, n.1-3, p.107-133, 2003.

YAMAMOTO, P. T.; PINTO, A. S. de; PAIVA, P. E. B.; GRAVENA, S. Selectivity of pesticides to natural enemies of citrus pests. **Laranja**, Cordeirópolis, v.13, n.2, p.709-755. 1992.

MIX
Papier aus verantwortungsvollen Quellen
Paper from responsible sources
FSC® C105338

Printed by Books on Demand GmbH, Norderstedt / Germany